t^{to}
4

THE QUEEN'S HIDDEN GARDEN

THE QUEEN'S HIDDEN GARDEN

*Buckingham Palace's
Treasury of Wild Plants*

Text by
DAVID BELLAMY

Botanical paintings by
MARJORIE LYON

DAVID & CHARLES
NEWTON ABBOT   LONDON

For Neil Duncan

British Library Cataloguing in Publication Data

Bellamy, David
 The Queen's hidden garden.
 1. Urban flora—England—London
 2. Buckingham Palace Gardens (London, England)
 I. Title   II. Lyon, Marjorie
 581.9421'32   SB466.G8B82

 ISBN 0-7153-8590-9

Typeset by MS Filmsetting Ltd, Frome, Somerset
and printed in The Netherlands
by Royal Smeets Offset BV, Weert
for David & Charles (Publishers) Limited
Brunel House Newton Abbot Devon

CONTENTS

Privet
Ligustrum vulgare L.

Ash
Fraxinus excelsior L.

(plate 1)

PREFACE

'They're changing guard at Buckingham Palace', and anyone who has ever visited the Great Metropolis of London as a tourist must have at least tried to press his or her face against the railings to enjoy the spectacle and wait, probably in vain like Christopher Robin and Alice, to catch a glimpse of the reigning monarch. Many will also have walked the great perimeter wall with its cheval-de-frise and wondered among other things whether the cruel spikes revolve and what it is really like beneath the graceful branches of the Royal trees which raise their canopies of gold or green at different seasons of the year.

Fewer people—but still a considerable number, for the Queen invites some 27,000 guests each year—will have responded to the Royal Command to take tea on the lawn, complete with cucumber sandwiches, scones, cream and a perfect infusion of *Camelia tea* with a hint of bergamot orange, all laced with the excitement of a walk around the Royal back garden. No wonder the roses both red and white do so well with a mulch from all those tea-leaves.

I must confess that a Royal Garden Party which includes at least 8,999 other pairs of tramping feet isn't perhaps the best time to seek out the botanical treasures of those hallowed acres, but with diligence you can try. In fact over the years, many botanists have been amongst the ranks of party goers and some have taken note—but never on the backs of their invitation cards—of the common plants which grow in profusion amongst the cultivated splendours of the Royal Garden. Please, I am not rebuking the gardeners—Messrs Wyness, Humphrey, Brown, Stirling, Carroux, Osborne, Cole and Nutbeam, to name but a few—of the last 140 years. They and their staff have, over the years, done Trojan work to meet the challenge not only of the chattering feet but also of the ravages of war, smog, a warm urban microclimate which receives a mere 20in of rain a year, and the weeds themselves.

To augment these somewhat clandestine records, experts and colleagues of the South London Entomological and Natural History Society were granted Royal Permission to investigate the natural history of the Palace Grounds. Over a period of four years they presented their visiting cards at the Garden Gate and, complete with vasculums, binoculars, lenses, floras, faunas, sweep-nets, light-traps and all the other paraphernalia of their various callings, revealed amongst many other things what is best described as 'treasury of common plants'.

The flora of Central London compiled over many more years, though to a large extent by the same people, listed only 475 species of wild and naturalised plants found growing amongst the offices, factories, homes, roads, parks and gardens of this urban 12,500 acres. To date 254 plants which fit the same description 'wild and naturalised' have been seen within the Palace Gardens—a treasurehouse indeed.

A walk around the gardens can only make this appear even more surprising for, of the 48.59 acres, 10 are covered by buildings, 20 by well-clipped greensward, and 3.63 by a serpentine lake. However, a closer look—and if I may be so bold, a poke around the borders, corners and compost heaps—reveals a right Royal load of weeds. To put it in perspective, with no volition of her own Her Majesty the Queen has a Nature Reserve right in her own back yard.

A Nature Reserve? But they are only weeds! Well, what exactly is the definition of a weed? 'Weeds are plants which are found growing in the wrong places', is the glib reply. In strict phytogeographical (the branch of botany which deals with the gross distribution of plants) parlance, that definition would include the vast majority of the plants which have been propagated and tended with such loving care by the Royal gardeners over the centuries. The truth is that many of what we like to call our English garden favourites have been culled from foreign lands and naturalised to grace our gardens. Even those which did have their origins in Britain are not usually grown within the spot or niche ordained by nature. The Queen's Rhododendrons and Azaleas—130-plus species, sub-species, varieties and cultivars no less—with their origins in Asia, China and North America; Honeysuckles from the Himalayas; Southern Beech from New Zealand; Bistorts from Russia and even more homely things such as Dorset Heath, Strawberry Tree, Box and Yew, though native to these islands, would never have been found growing naturally on the Royal site.

Turning to another definition of weeds we find that, in the agri-business, a weed is any plant which competes with the crop for which the ground has been prepared. This in many ways would appear a more apposite and meaningful definition. However, we shall learn that the soil upon which our crops depend was put there in the first place in part through the activity of natural (weed) communities and, what is more, is held there and maintained in good heart by the remnants of those plant communities—hedges, copses, wetlands, Sites of Special Scientific Interest (SSSIs) and the like—if only we would let them be. So even this definition should stick in the throat of every real farmer.

Weeds are simply plants which don't need pampering, they have the ability to take up each and every opportunity which we put their way, and that they do so is as much our fault as theirs. For the moment we will comfort ourselves with the knowledge that all weeds aren't all bad, and with the definition which has been attributed to Ralph Waldo Emerson: 'A weed is a plant whose virtues have not yet been discovered.'

Before I continue this story of glittering success, that is for the plants concerned, I must thank many people. First and foremost Her Majesty Queen Elizabeth II, who graciously gave her consent not only to the publication of this book, but that we through the good offices of her staff have had access to the gardens of Buckingham Palace. Then to my co-authoress Marjorie Lyon, though her illustrations speak of her talents much more eloquently than I can, demonstrating as they do the true beauty of even the most maligned and malignant weed. Then to the officers and members of the South London Entomological and Natural History Society who, through the publication of their transactions, put so much

8

information on permanent record. My special thanks are also due to James Greig, Paul Moxey, Robert Scaife, Judith Turner, Martyn Kelly, whose palynological knowledge and unpublished works have been of great value, Peter Sell of the Botany School, Cambridge, for his expert help with Hawkweeds, Caroline Rockman for her help with the Treasury of Common Plants and Meriel Steel who not only typed the manuscript but also helps her husband farm part of Britain to perfection.

As to my sources of reference, they are many and are listed on page 217, not only to beg forgiveness for my abject plagiarism, but to introduce my readers to the real zeal of the real experts in each subject. I must not forget my undergraduate and postgraduate students at Durham University, whose brains I have unashamedly picked in essay and tutorial for over twenty years.

Finally I must say a thank you to the plants themselves for being there and for making all our lives possible; and above all for making my own life one of wonder and fascination, spent in great part trying to discover what it is that makes them tick. Now for the weeds!

ARTIST'S ACKNOWLEDGEMENTS

I should like particularly to thank Her Majesty the Queen for her gracious permission to visit the garden of Buckingham Palace. I wish also to thank Mr Fred Kemp, Acting Head Gardener at Buckingham Palace, who was extremely kind. For Mr Kemp I illustrated *Galinsoga parviflora* which, he informed me, is the most prevalent weed in the garden.

To David Bellamy I feel especially indebted for taking such an enthusiastic interest in my work. I am greatly honoured.

Thanks are due to several people for botanical information: Margaret Hanbury, Anthony Huxley and Dr Alan Leslie of the RHS Garden, Wisley, were very kind indeed; Dr Brinsley Burbridge and the staff at the Royal Botanic Garden, Edinburgh; John Forsyth, Adviser in Environmental Education for the Border Region, for his unfailing assistance and expertise; Marion Forsyth; Mr and Mrs Grant Roger; Shirley Haward; Lord Binning; Mrs Anna Younger; Michael Braithwaite and Dr Neil Duncan.

I am very grateful to Dr Pat Monaghan who provided me with such excellent specialist books and access to the Department of Zoology, Glasgow University; for the expert verification of my moth drawings thanks to Dr R. Crowson.

My heartfelt gratitude to Elizabeth Lyon, Alistair Allport and Ronald Gunn for their invaluable comment and advice about all matters artistic as well as vital information about plants. For a great deal of support I must acknowledge the kindness of Mr James McColm, Mrs S. Forrester and Mr Andrew Wilson. My very special thanks to Malcolm Forrester without whose continuous help and encouragement I would not have embarked on a painting career. Most of all, I am extremely grateful to Joan and Bill Fraser for their unstinting generosity and practical assistance in connection with this book.

Bugle
Ajuga reptans L.

Black Horehound
Ballota nigra L.
ssp. *foetida* Hayek

Gipsywort
Lycopus europaeus L.

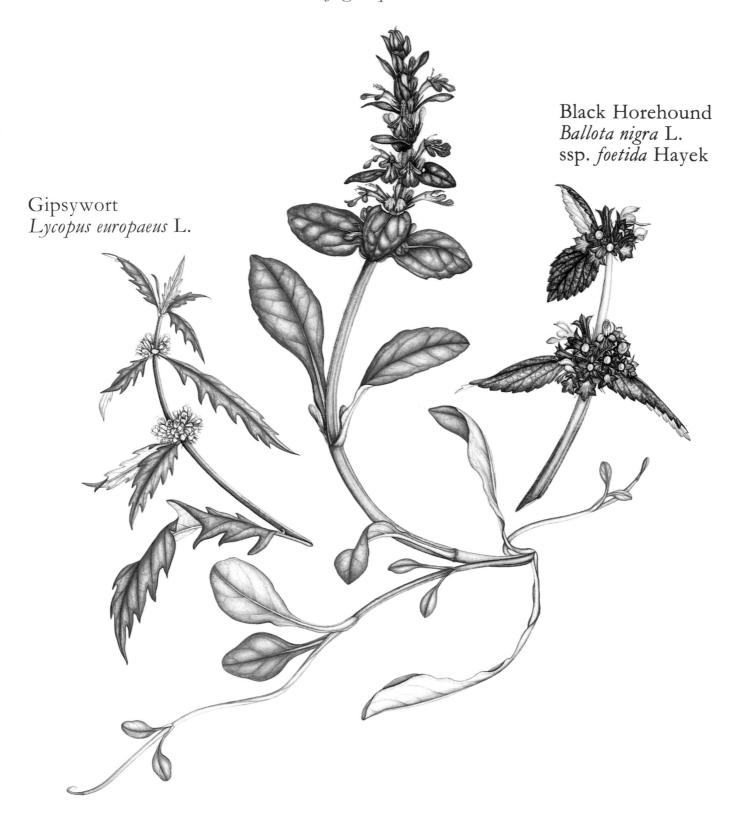

(plate 2)

A WEED FOR ALL REASONS

Having decided that weeds are vegetable opportunists—plants which may nip in and take up the potential of any mismanaged habitat which may become available—we can now ask what it is that a plant needs in order to be a success in the world of weeds.

First and foremost, it must be able to move—no mean feat for an organism with roots; that is, unless the plant produces an abundance of seeds, each of which possesses some method of transportation. Seed release may be no more of a problem than falling off a log, but getting those seeds away from the immediate sphere of influence of Mum who, in the plant world, is always a shady character, is always a problem.

Fortunately, you don't need a leaning Tower of Pisa to prove that, despite the fact that gravity works equally on all objects, in the thin air which surrounds our planet, a feather will fall more slowly than lead shot of the same weight. If we lived in a vacuum unsupported by anything else—and I must say that some people's attitude towards the environment and wildlife makes me believe that they think they do—then all objects regardless of mass and shape, would fall at the same rate. However, thanks to the atmospheric envelope (which is maintained in a pure, life-supporting state by the activity of all the other plants and animals with which we share this planet), the findings of Archimedes must be applied when considering the rate of fall of any object: any object displaces its own volume of air and every person displaces his or her own volume of bath water and natural habitat.

Terminal velocity always sounds a pretty final thing, and it is, for it is the maximum speed an object reaches when falling freely under the force of gravity through the atmosphere. Certain seeds like the Coconut, or indeed the Double Coconut or Coco de Mer, the largest seed in the world, are so large and heavy—13lb for a really big one—that Archimedes' principle doesn't get a look in, and they fall straight down, leaves permitting. Anyone who gets in the way will receive a short sharp lesson on both the force of gravity and terminal velocity.

The Coconut has cashed in on the fact that it can float, survive a long sea journey, and germinate and thrive once washed up on some sandy shore. It thus has what it needs to become a weed. The Coco de Mer, on the other hand, floats in vain; for it can only germinate and thrive within the shade of its own forest, and so it has remained a monstrous rarity, marooned in one valley on one island in the Seychelles.

Apples, of course, come into the category of fast fallers, and it is said that Isaac

Newton was sitting under an apple tree when the principia of science descended on his head.

It must be remembered that although most of our commercial fruits have been selected and bred to be heavy and juicy, the vast majority of plants with heavy seeds and fruits have their natural home in tropical forests. Under the dense shade of the continuous canopy there is little or no advantage in a seed being transported away from the parent tree. In such situations it is however of great importance for the seeds to contain as large a food store as possible, sufficient to give the seedlings a good start to life down in the darkness of the forest floor. The cavity in your avocado, which overflows with prawns and vinaigrette, represents half of the size of the seed, much of which is packed with food for the well-being of the embryo. The succulent flesh you eat is a shock absorber which cushions the seed as it hits the branches and the ground, and a good handful of damp potting compost to help get the seedling rooted.

Descending from the monster cannonball fruits and seeds to the other end of the size range, we come to the 'dust' seeds, which vie with pollen grains and the spores of ferns and mosses and the like when it comes to floating in thin air. The Orchids are perhaps the most spectacular of the dust-seed producers, for a single capsule of one of our common Orchids may produce millions of seeds.

Although the weight of such a seed may be as little as $1/50000z$, so that you would get more than 90 million in 1lb, they weigh much more than the volume of air they displace, and so they do tend to fall vertically down. However, their terminal velocities are only around 4in per second, and so they are very susceptible to the eddies and convection currents which are part of the air mass even on the stillest day. So dust seeds can remain suspended in the air and float for enormous distances. Like pollen grains and spores they may also be buoyed up even into the stratosphere where they may drift from continent to continent, that is if they are not retrieved by high-flying Jacques Piccards and the odd meteorological balloon.

Just as the heavyweight seeds are mainly produced by plants of tropical forests, so too are the dust seeds. The former are produced by trees and tall shrubs, the latter by epiphytes, which grow attached mainly to the high branches of those trees. Like most of the world's ferns, the majority of the world's orchids live this rather precarious high-rise hanging-about existence. Their multi-millions of spores and dust seeds rise up in the thermals above the forest canopy to be carried to branches new, where they can germinate and live it up in the well-lit canopy.

The only real problem faced by such aerialists is that they can't pack much in the way of food for the journey and the hopefully happy landing into such a small seed. Therefore the vast majority of these dust-seed producers rely on very specialised methods of seedling nutrition. Some are total parasites, which means that the seed must fall onto, or close to, its chosen host, and once there feed directly from it. The vast majority are, however, known to be dependant on specific fungi being present in the habitat. These fungi enter into close association with the seedling, keeping it supplied with nutrients through this crucial stage—a sort of intensive baby care unit.

So although dust seeds may be the tops when it comes to long-distance transport, they have great limitations as regards the type of habitat they can colonise. If the right nurse fungus is not present they will not survive. This does not, however, exclude the dust-seed producers from the ranks of the opportunists; and I have on a number of occasions seen orchids in their thousands colonising newly cleared ground and even industrial waste. However, their specialised nutrition does exclude them from the ranks of the true weeds.

Certain plants—and the list includes some of the world's most successful weeds—have overcome the limitations of floatability and size by adorning their seeds with Archimedean parachutes of one sort or another; that is structures and devices which increase surface area and hence drag, without adding too much to the weight. Perusal of any recent edition of *Flora of the British Isles* will show you that the largest family of flower-bearing plants in the world is the Daisy Family, which boasts in excess of 14,000 species. In the description of the family, it says: 'fruit an achene [a small one-seeded nut] crowned by the pappus [parachute]'. No wonder that out of the full list of the Queen's Weeds, no less than forty-two (17 per cent) are members of this family.

Detailed studies of seventeen very successful members of the Daisy Family have been made, and have shown a strict relationship between that all-important terminal velocity and the ratio of both the weight and the diameter of the achene and its parachute. The greater the weight and girth of the fruit, the faster they fall, but the greater the mass and especially the diameter of the parachute the slower they fall, and the better the chance of being dispersed by the wind. However, the actual structure of the parachute also plays a part, for the very openwork pappus possessed by the Cat's Ear means that its fruit falls almost twice as fast as that of the Blue Fleabane, despite the fact that there is the same ratio of chute to fruit diameter.

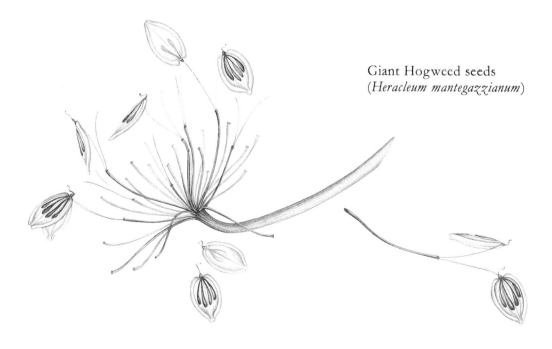

Giant Hogweed seeds
(*Heracleum mantegazzianum*)

Great Water-grass
Glyceria maxima
(Hartman) Holmberg

Sweet Vernal-grass
Anthoxanthum odoratum L.

(plate 3)

Any human parachutist will tell you of the perils of falling onto water, that is unless you are trained and equipped for such an eventuality. The same is true for the plants. Some seeds have absorbent parachutes which collapse on contact with water, while others like those of the Dandelion and Coltsfoot are non-wettable, and so once ditched they continue on their way, sailing at tremendous speeds. Coltsfoot seeds have been clocked in at 44yd a minute, that is $1\frac{1}{2}$ miles an hour—not bad for a little weed.

It must be remembered that the production of any structure, however flimsy, represents the investment of both energy and raw materials by the plant. Wild plants and animals don't have credit cards or banks to tide them over impecunious times. If they invest in a structure or a stratagem, that investment must pay off. If it doesn't they will lose out and may well go to the wall of extinction. There is, however, no getting away from the fact that the energy invested by generations of the Daisy Family in pappus parachutes has paid back many million-fold, floating their seeds of success across an ever more inviting weed world.

However perfect your parachute, there is one overriding factor which determines the success of the whole operation, and that is the height from which it is released. Just watch your local crop of Dandelions, and you will see the hollow stalk of this pre-eminent weed rise to the occasion. As the fruits ripen, it elongates, clocking up the extra centimetres which may make all the difference. I wonder whether non-wettable seeds have got anything to do with its oft-quoted diuretic properties.

Parachute fruits are not of course the sole prerogative of the Daisy Family, and other examples spring to mind at various seasons of the year. The Buttercup Family tops the aerial act when the Clematis comes into fruit. The long silvery white-plumed persistent styles, one to each fruit, crowd our hedgerows and their flight paths. No wonder another name for this plant is Traveller's Joy.

Poplars can be unpopular because their silky white lint cascades all over the lawn. The most prolific member of this family is the Cottonwood of North America, which produces a veritable white-out over vast areas. Without the parachutes of the Cotton-plant *sensu stricto*, the whole course of history of a large slice of the world which was once part of the British Empire would have been very different. It seems somewhat ironical that it was these snow-white structures, evolved to have the freedom of the winds, that brought slavery to so many people. The scientists and plant breeders also did their best to stop the fruits from escaping, that is from falling off the plants. They did not want to clip their wings because these were made of precious cotton, but they did want to fix them to the plant until the time of harvest. What is more, they managed to do just that, and each advance in cotton-boll technology sent a quiver through the London Stock Exchange.

When it comes to London proper, the most famous of all its flying fruits belongs to a member of the Fuchsia Family or, to give it its proper name, the Onagraceae. This beautiful plant blew in from the fields during World War II. Its seeds sallied forth up our shattered alleys and streets covering the bombed sites with a veil of purple pink, hiding the scars of human greed and stupidity. Its name, the Rosebay

Willow-herb (for willow trees produce similar plumed fruits) or Fireweed; for in nature it heals the scars after forest fires.

It is of more than passing interest that of the thirteen species of Willow-herb found growing in Britain today, two are introductions. One was introduced from North America only in 1891, but is now found throughout Britain even in the Outer Hebrides. The other came from New Zealand, but is now well-established on moist stony ground in the mountains of England, Scotland, Wales and Ireland.

From parachutes to wings—and with the current explosion of paragliding, hang-gliding and microliting amongst the human populace it shouldn't be too difficult to understand that there is a whole evolutionary series between the two. In fact the only thing we find missing in the plant kingdom is true powered flight. Rocket ballistics are commonplace, especially down in the dank world of the mushrooms and toadstools. However, the weeds do get in on the act in their own explosive ways.

Exploding Policeman's Helmets may seem like a figment of *Goon Show* imagination. They can however be seen, felt and heard on river banks and shady places throughout England in August and September. The Policeman's Helmet— the plant that is—was introduced from the Himalayas during the time of the British Raj, and is today fully naturalised and doing very well in an ever-increasing number of places. The flower looks not unlike the headgear of a member of the Keystone Cops, and the fruit, once ripe, looks not unlike one of their trusty truncheons. The fruit, which is a capsule, dries as it ripens and the sections of which it is made try to shrink, so that the whole is put under considerable tension. Along comes a wandering bumble bee, botanist or even a breath of wind, and the whole thing cracks up, flinging the ripe seeds out in all directions. The effectiveness of the explosive dispersal came to me when I was quite young. I remember collecting seeds from the banks of the River Mole in Surrey, and planting them in my garden. The first year went well, all the neighbours came in and admired the sweet-smelling flowers, and popped the exploding fruits. Next year it wasn't so good, for a great army of would-be Policeman's Helmets appeared marching across the lawn, much to my Dad's annoyance.

From an exploding import to an exploding export—Gorse or rather Gorses; for we have three which grow commonly in Britain, and unfortunately now are found growing as very noxious weeds in far-flung corners of the ex-Empire. Gorses are members of the Pea Family, and they were taken overseas by emigrant farmers to remind them of home and provide a spiny fast-growing fence around the new homestead. Unfortunately, in amongst the spines they produce special sorts of Pea (sorry, Gorse) pods, which come under tension as they dry and explode when ripe, flinging out the seeds. Flying seeds are only part of the problem; the root cause is that like many members of this family, Gorse roots produce nodules which are infested with bacteria which fix nitrogen from the atmosphere and turn it into nitrate fertilizer. So what with explosive fruits and self-fertilizing roots, Gorse didn't stick to the edges of the field, and instead of protecting the homestead invaded it, popping about all over the place.

Care must however be taken, not only when it comes to moving plants about the world, but also when it comes to listening to them, for not all loud pops signify such active dispersal. Take the Bladder Senna for example, a handsome shrub which was introduced from the Mediterranean and is now well at home around London. Its main fascination is its large silvery 'pea' pods which are shaped like barrage balloons. In the Language of Flowers, it was said to denote Idle Pleasure, and I must agree with many people that it is great fun to pop the fruits. It is, however, a truly idle pleasure for the seeds are not released by the explosion. Indeed, if the fruit is left intact on the plant, it just withers without even splitting its own sides.

Anyone who has watched with envy an exponent of hang-gliding making his way gracefully through the thermals, will realise that the best design for a glider is to have the weight placed centrally within the aerofoil surface. They will also realise that by slightly shifting and angling the central mass, the flimsy craft may be flown to perfection or crashed with conviction. The same is true of seeds, and of all our native plants the Elm and the Ash come closest to such aerodynamic perfection, with the Docks not far behind, although they have the limitations of a lower launch-pad. If you collect some of these tree fruits and fly them out of your bedroom window, you will find that, as they ripen and wrinkle, all sorts of aerobatics become possible.

The biggest and best of these gliding fruits are again found in the tropical forest, belonging to the Lianas which climb the trees and so keep up in the sunlight. For them a graceful undulating glide is probably the only way to carry a moderate-sized survival pack to branches new.

There is little doubt that of all the winged seeds, the best known both in Britain and North America belong to the Maples. These not only glide, but because of their weird lop-sided shape rotate as they go and that rotation induces lift, prolonging the flight. Although one sort of Maple is native to Britain, two others have been introduced and are grown quite commonly.

Autogyro fruits of Sycamore (*Acer pseudoplantanus*)

Yellow Flag
Iris pseudacorus L.

(plate 4)

In the native Common Maple the wings of the fruits spread horizontally, whereas those of both the Norway Maple and the Sycamore spread at an acute angle, and in the latter the tips may be incurved. Perhaps this slight incurving is an anti-stall device which along with hairless shiny fruits has given it a slight edge over the others, for since its introduction from Europe in the fifteenth century it has taken over Britain and is now our commonest weed tree. The curved wings of success—but there is little doubt that if man hadn't brought it to Britain, Sycamore would not have been able to make the channel crossing all by itself.

So we come to the last really important agent of seed dispersal—animals, including us. Here again we find an enormous investment of energy, not to say ingenuity, on the part of the plants in annexing the help of the more mobile members of the living kingdom. Some of the animals, as we shall see, take up the seeds on their own volition; but others have the seeds as it were, thrust upon them. The beautiful colours and aromas of many fruits are not put there to satisfy fads of the judges at our autumnal horticultural shows, nor the preservers and picklers of the Women's Institutes, nor the devices and desires of the Constance Spryers, bless their hearts. They are there to attract the mobile hordes, and the more mobile, the better.

Birds come in most shapes and sizes, from the aerodynamically impossible buzzers like the Humming Birds, to the enormous, graceful flappers like the Buzzards. The former expend large amounts of energy just keeping up, and so have to refuel constantly in flight; the latter do it the easy way, and know more about thermals than all the blanket makers in the world. When it comes to fruit eating, birds however come in two basic types. Some have hard gizzards which they fill with stones, making them so hard that they can grind the fruits up, seeds and all, to extract all the goodness stored within. Turkeys say 'gobble, gobble', and do exactly that, for they are seed crushers par excellence, as is one of the most colourful visitors to our gardens, the Hawfinch, who has the added advantage of a beak which can itself crack open cherry stones. Then there are those with soft gizzards; and although they can eat and digest the soft parts of the fruit, the hard seeds pass right through, ready for planting—a fact which may give the bird the pip, but is just what the plant invested all that energy for. Thrushes have soft gizzards, and hence spread the seeds; and it is perhaps more than of passing interest that the Latin name for these birds is *Turdus*, which would make an appropriate name for the potting compost they produce.

Sometimes the two types of birds interact, and it is not unusual to find Thrushes, Redwings and Fieldfares gorging themselves on rosehips and enjoying all the vitamin E. Meanwhile the Hawfinches wait in the wings feeding on hawthorn haws before they swoop in to scoop up the bird lime full of undigested rose seeds. This habit of dropping waste full of seeds often leads to the clumping of seedlings of one or a number of species, through growing in such close-packed proximity it is rare for more than one to survive. However, that one will be helped on its way by the nutrients and organic matter contained in the dung and in the others.

Some birds don't even bother to swallow the seeds, but store them in their beaks, ejecting pellets as they go. Again these pelletised seeds (there's nothing new in gardening) have the advantage of added nutrient, but again often grow in clumps. Other birds are more fastidious, and wipe their beaks on anything which comes in handy, like the branch of a tree. If the seeds are sticky—as they often are when all gooed up with fruit juice—they may stick to the branch, or the bird may have so much difficulty getting rid of them that the seeds are wedged into cracks in the bark. It is said that this is the main way in which the seeds of Mistletoe are transferred from tree to tree.

Clumps of seeds and seedlings may likewise result from the recycling habits of other types of animals. It was recorded way back in 1911 that a cow grazing in a weedy field could in a day consume as many as 89,000 Plantain seeds and 564,000 seeds of Chamomile. What is more—and the researcher deserves much more than a pat on the head—it was found that 85,000 of the former and 198,000 of the latter avoided the repetitions of chewing the cud and the churning attention of the rumen, reticulum, omasum and abomasum stomachs to emerge still recognisable at the other end. Recognisable, but not unscathed, for only 58 per cent of the Plantain and 27 per cent of the Chamomile seeds were in a viable state. Add to that the problem of germination and establishment in large dollops of dung which includes many substances in concentrations which will scorch the seedlings, and you will see that life in a cowfield isn't all a bed of roses even for the most adaptable weed. Study of overstocked fields will also show that too much dung can lead to the death of the whole sward and to little burnt bald patches or long zigzag bald trails, the amplitude and pitch of the zigzag telling you how fast the cow was

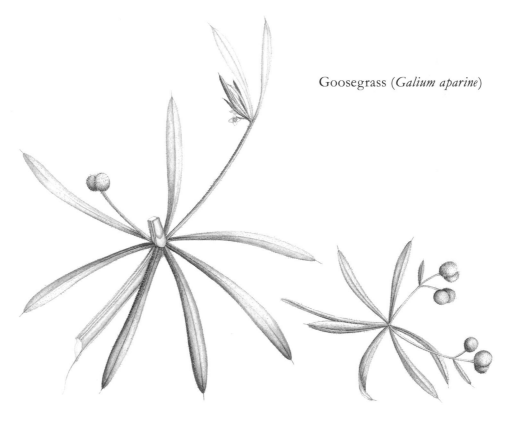

Goosegrass (*Galium aparine*)

moving about its business. Distasteful as the whole subject may seem, it is of great importance both to plants and to modern dairy farmers; and demonstrates the prodigious weed-seed production and the problems of even a highly bred vegetarian—facts to be ruminated upon.

Not only are seeds carried about in the inside of animals, but also on the outside, and to accomplish this they are armed with all sorts of adhesive structures. All one need do to find out is to take a walk through a piece of wasteland and then count the seeds and other things adhering to your socks. How's this for the result of a short stroll across an untidy corner of 'Happy' Hampstead Heath? First, some 273 Sticky Willies, Goosegrass, Cleavers or Hairif. All are one and the same plant whose seeds, leaves and stems are covered with a multitude of tiny hooked hairs— nature's own Velcro, and just as efficient. By classification a very successful member of the Madder Family, there's nothing makes you madder than trying to get bits of it off your clothes. There's also Burdock, a member of the Daisy Family, whose seed parachutes are stiff and covered with scratchy hairs in several clinging rows. To be exact there are two sorts present, 193 Great Burdock and 18 Lesser Burdock, I think; well at least 18 of them are brownish compared to the others. Then there are some impossible seeds—amazingly annoying things about $\frac{1}{8}$in in diameter, each one densely covered with stiff, hooked white bristles. Seventy-six of them up to this point, and I am giving up. The rest can stay put between the plain and purl, and blow the size of the future population of Enchanter's Nightshade.

It isn't only the large animals like us which mobilise the seeds. Smaller mammals also do more than their fair share of seed moving, and both squirrels and mice may hoard many hundreds of seeds and fruits for winter with some bizarre effects in the spring if they forget where they hid them. Sometimes agents move seeds enormous distances. We do not as yet understand the exact mechanisms by which migrating birds find their way across the world so that they can return to grace our gardens with their spring, summer, autumn or winter presence, bringing with them seeds on mud-caked feet or fixed upon their feathers.

Insects, and especially ants, are big operators in the seed removal business; and although the latter can usually only manage one at a time, a nest of ants makes short work of even the largest store. The plants themselves cash in on the fact by providing oil stores as part of the seeds specially for the ants to eat, for then they leave the seeds to germinate. In other cases the relationship is a much simpler one, whereby the ants carry the seeds to their nest to consume them in their entirety. However, some always get lost en route due to insurmountable obstacles or battles with other ants and, forgotten, may germinate in an advantageous spot for the plants. Then there are the ants which store seeds in special underground chambers in their nests, ready to be fed to their young. If these seeds germinate while in store, the workers will laboriously carry them out to be deposited just beyond the edge of the nest-platform where they will germinate in what is well-fertilized soil.

So it is that wherever we care to look in the world of plants we will see adaptations which appear ideally suited for the weedy habit and habitat. One cogent question must however be asked. Have they in fact evolved in relation to the opportunities created by agricultural man; has there been sufficient time in the 15,000 years during which agriculture has been practised? The answer is two-faced, rather than two-fold. Many of the adaptations could have evolved in relation to natural catastrophies—hurricane, fire, flood or overgrazing—then our weeds would have been there, pre-equipped as it were, waiting for the opportunity. However, in the case of certain weeds, studies have shown modifications which can only be due to rapid evolution under agricultural conditions.

Take, for example, the case of Gold of Pleasure, a weed of flax and linseed crops. Investigations have revealed the presence of several races which ensure that they stay with the crop seed as it passes through the winnowing machine. Rapid evolution indeed with relation to a machine which was designed less than 400 years ago to sort seeds on size, density and structure, and a real headache for both farmers, agricultural engineers and seed merchants alike. For whether we agree about the rapid evolution or not, there is no disputing the fact that the farming community is the weed's best friend.

Despite the fact that agricultural departments the world over spend multi-millions of dollars trying to combat the problem, they have so far not begun to beat the top-class weeds. Even at the allowed level of infestation of grass seeds— 1 per cent of weeds by weight—each square metre of newly sown sward would receive 185 seeds of Chickweed, 92 of Ribwort Plantain, 74 of Yorkshire Fog and 12 of Curled Dock. Moreover, all the energy and expense which the farmer or gardener puts into preparing his soil for his crop is also lavished on the weeds, the majority being drilled into good ground to germinate under optimum conditions of tilth and fertilizer application. What is even more disturbing is that many, like the Pineapple Weed, sop the fertilizer up, preferentially leaving the crop plants to fend for themselves. Like all seeds, those of the weeds have problems; but many of these are overcome by the farmer, kind man that he is.

Unfortunately for weed seeds, not all of them have well-drilled farmers or gardeners to help them on their way; many just breeze in by themselves. Another name for a weed is an 'alien', and the habitats many of these seeds find for themselves are pretty alien places—open ground with little or no shelter from the vagaries of the weather for instance. We are foolish enough to talk about a temperate climate here in Britain, temperate meaning unchanging. Plants have to stay out of doors the year round, and in reality the weed seeds face an alien environment in which temperature and humidity can fluctuate wildly all the time, and they face it with no shelter at all. Bone dry one minute, soaked in distilled water the next; open to the full glare of the sun, then plunged into pitch darkness—for wind doesn't only blow seed, it also blows the soil about.

Then there is the microtopographical variance. From 5ft plus, a well-prepared seed bed may look perfectly uniform to the eyes of a gardener. However, down at seed level the soil particles take on the dimensions of a mountain range. I have a times 30 field microscope with a light in the end, and I get hours of fascination flat on my tummy viewing the soil landscapes. Big bits, small bits, soft bits, hard bits—a jumbled world of all shapes, sizes and textures with nooks and crannies, holes, caves, peaks and crevasses. I am about thirty times the height of a small apple, so

Common Mullein
Verbascum thapsus L.

Sharp-leaved Fluellen
Kickxia elatine (L.) Dum.

(plate 5)

imagine what the soil would look like, at a comparable scale, to an apple pip or even a dust seed. John Innes No 1 begins to take on the aspect of the Himalayas. No wonder many seeds fall or get washed down amongst the soil particles, lost from view and into a situation in which they may lie dormant for a considerable time.

Dormancy is an important part of the life cycle of many organisms, be it the sleep of a night, a winter or a drought. In animals it is characterised by a snore and a store of energy which is slowly used up during the period of inactivity or complete hibernation. Sleep is a regular requirement, and probably a very necessary one, in warm-blooded animals with a high rate of living. The 'sleep' movements of the Wood Sorrel which folds its leaflets down at night and the midday flower-nap of Jack-go-to-bed-at-noon relate to the daily cycle in which photosynthesis predominates in the light and respiration in the dark. This cycle is further complicated by the fact that when actively making sugar by photosynthesis, most plants have to have their stomata (leaf pores) open and hence lose water.

(left) Ribwort Plaintain (*Plantago lanceolata*), (centre) Rat's-tail Plantain (*Plantago major*), (right) Hoary Plantain (*Plantago media*)

The fact that most of our cold-blooded animals are diurnal—that is they live it up in the day—relates to the fluctuations of temperature; they need warming up. The fact that most of our mammals are nocturnal—live it up in the night—relates to predation; it is easier to hide in the dark. Both facts help to regularise their patterns of activity.

The main long periods of dormancy both in plants and animals, however, relate to the seasonal development of adverse environmental conditions, be it the frosts of winter or the droughts of summer. Such hibernation is found in just about every group of plants and animals which live on or close to the land, from the single-celled plant/animals found in every drop of water through the lung fish and amphibians which inhabit the edge of lakes, ponds and streams to our hedgehogs who rolled up regularly in our country lanes and gardens with no real enemies until the motor car came on the scene. At one time it was thought that swallows hibernated in the mud at the edge of farm ponds and re-emerged each spring. But, such mobile organisms have no need for hibernation, for if they cannot find a livelihood through cold or drought, they can simply migrate to warmer or damper climes.

When it comes to plants, we don't begin to understand the full complex of factors which allow their seeds to enter and then to break dormancy, and the more research that is carried out the more methods or stratagems seem to come to light. First and foremost there is temperature. Certain seeds germinate—and come to that, bulbs, corms, rhizomes and buds burst into life—when the temperature reaches a certain threshold level. An effective, but dodgy procedure, because any gardener will know the problems of an early warm spell followed by late frosts. Similarly, in a seasonally dry environment the onset of the first rains may bring dormancy to an end with catastrophic results if it is a mere passing shower and doesn't soak the soil.

A more standard time-switch is provided by the solar clock, and day length or rather the ratio of daylight to night-dark controls various activities of many plants. It is, in fact, one reason why we have both wild and garden plants which bloom at different times of the year. The flowering of Chrysanthemums which are turned on by the shorter days of the autumn equinox can be held back by a little light in the greenhouse, so that they think it is still the long days of summer. Thus they can be tricked into flowering for Christmas. The same is true of the long-day plants which flower in summer, although it is more difficult to darken your greenhouse and hold them back. But of course it is easy to bring at least some of them forward. Just as these plants regulate their flowering by the solar clock, others regulate their dormancy in similar manner. However, such plants cannot cash in on either the promise of an early season or the potential of a late one, but they should in most normal years avoid the worst of frosts and droughts.

If the seeds of a plant are going to respond to the solar clock, they must of course be in a position to 'see' the light, and the deeper they fall or get washed into the soil, the less chance there is of that happening unless they are in soil which will be disturbed by ploughshare, harrow or spade. So it is that seeds of many of our worst weeds stay dormant until deliberately exposed to light. What is more, their light-

Ivy
Hedera helix L.

(plate 6)

sensitive systems must be very sophisticated for just any old light will not work, it must be of the correct wavelength.

Other seeds benefit from abrasion during digging or ploughing, for their coats must be chipped or damaged in some way before they are able to germinate. Such mechanical damage opens up a chink which can let in light, oxygen or water, or let out accumulations of carbon dioxide or some other product of the dormant seed which may be a specific germination inhibitor produced by the plant. The Wild Oat is a good example so remember, should you ever want to sow some on purpose, chip them first.

Certain seeds need frosting before germination can be triggered, and the whole science of vernalisation (springification) and the development of spring-sown wheat depends on the knowledge that a deep freeze is ample substitute even for a Russian winter. This is, of course, a really crafty trick as far as the plant is concerned; for if after seed-fall an Indian summer provides the right conditions for germination, the seedlings will not pop up in vain, they will remain dormant until winter has passed.

Later in the year, temperature fluctuations can trigger the germination of some seeds, and here the receptor mechanisms are very sensitive indeed. Why, in the Poppy which was once probably our commonest weed of arable land, should the vast majority of the seeds germinate only when subjected to a daily fluctuation of temperature of between 10 and 30 degrees C (50–86 degrees F), when 10 to 20 degrees C (50–68 degrees F) or 20 to 30 degrees C (68–86 degrees F) has no such mass effect? The fascinating thing is that a daily fluctuation in that range of a whole 20 degrees C (36 degrees F) is unlikely to occur even in bare cornfields until the April/May interface, when conditions are perfect to produce a beautiful or annoying carpet of red, depending of course on who is looking at it.

What is more, and perhaps strangest of all, weeds do not put all their seeds into one stratagem for breaking dormancy. Very often the seeds which are produced first, even on one flowering head, have a different wake-up requirement to those produced even a few days later. Fat Hen may produce as many as four different types of seed varying on black, brown, smooth and wrinkled. The brown seeds hardly enter a dormancy period, germinating soon after they have fallen. In contrast, the black seeds require chilling or a dollop of nitrate fertilizer before they can get cracking. Thus the brown seeds produce autumn plants which, if they manage to overwinter, produce very large plants in the second year with an abundance of flowers and seeds. The black seeds germinate in the spring, thus doing their best to extend their chances not only of survival, but of success.

When it comes to the breaking of the dormancy of seeds, there are so many fascinating facts and factors that have been and still wait to be discovered, that all I can do is whet your appetite. If you want to know more, either read the specialist accounts or try to find out for yourself. Don't look upon your weeds just as a challenge to your prowess as a gardener, but as a challenge to your understanding. Your garden is a laboratory in which you can begin to find out for yourself; and the more of their strategems you can understand, the easier it will become to accommodate the weeds into your garden life. Never rest on your laurels, and always remember that you can never win. Suffice it to say that, as far as the weeds are concerned, the end result is a seed bank lying dormant in each and every soil, each seed type and indeed each seed slightly different from the rest. Their particular adaptations have been programmed in by evolution through natural selection both long-term and rapid, and allow them to make use of every opportunity which you put on offer. It is frightening to think of all that store of 'knowledge' hidden underground, ready to outwit us as soon as we turn the soil.

All-seed (*Chenopodium polyspermum*)
and Fig-leaved Goosefoot (*Chenopodium ficifolium*)

As a botanist I am always being asked two questions: can plants talk to each other and do they feel? My answers are always the same. Yes, they can communicate certain facts to each other, to their progeny and to animals who visit them. However, it isn't a language like ours, composed of noises, words and sentences. It is a language of chemicals—detailed information which is passed around the plant, on to the next generation through the genes, or diffused out into the soil, water or air in which they live. In the same way it is evident that plants have senses just like us, although they have no need for the sophistication of hands, ears, eyes, noses and tongues to make them work. Why, we have already seen that seeds can sense light, time, temperature and abrasion—not exactly the same as sight and touch, I must agree, but not far off. Certainly the seeds of certain plants which can only germinate successfully in acid, neutral or alkaline soil, or in soil in which there is sufficient or not too much oxygen or carbon dioxide or some specific aromatic substance, are 'tasting' or 'smelling' the soil. The song about the lonely little petunia in the onion patch made popular the fact that certain chemicals released by dense crops of onions and leeks inhibits the germination and growth of weeds. It isn't just a recently realised phenomenon either, for the Incas of Peru planted certain sacred plants amongst their corn to keep away the weeds and it worked. Recent studies have shown that they do produce herbicides which inhibit successful germination of the weeds.

Sight, touch, smell, taste—the only human sense which seems to be missing from the seed bank is hearing. There are many instances on record of research which has claimed all sorts of effects of sound in promoting or inhibiting plant growth. Some experiments have been carried out with enough care and attention to warrant the conclusion that plants are affected by sound. However, as far as I can ascertain, none have proved any such effects on the seed and, until such time they do, we must conclude that seeds are as deaf as the proverbial posts into which their trunks may one day turn—that is of course if they can break dormancy.

There is, however, one sense which plants appear to have which we do not, namely the sense of gravity. Complex characters as we are, we perceive gravity by watching or feeling the effect it has on something else. Plants may well be able to sense gravity, their roots turning down and their shoots turning up. This is, in fact, how we know they feel it; for although botanists have striven to find something within the living cells of all roots and shoots which fall and thus trigger the gravity response, they have so far failed miserably. It is strange to consider that this may well be our limitation and not the plants', for if they really do enjoy a sense which we do not have, how can we begin to find out?

One thing is more than certain; we know that there is one major difference between the senses of plants and animals. The latter are based on a nervous system, and in the higher animals the impulses from that system are de-coded and translated by the brain. It is easy to demonstrate that plants don't have brains and nerves—no one has ever found them. However in the knowledge that both plants and animals are made up of living cells which, though different in gross structure, have many of the simpler sub-cellular structures and biochemical (living) pathways in common; that nerves are modified cells and that the nervous impulse is electrochemically induced; it would be a foolish scientist who said that plants have no feelings. So have a care or at least spare a thought when you next thrust your spade into that non-listening seed bank, and don't run away with the idea that the seeds won't know what you are doing.

The seed bank is both diverse and well stocked; in most soils there are more potential plants lying dormant in the soil than there are individual plants growing on it. Take for instance the Broadbalk Field at the Rothampsted Agricultural Research Station in England. Its agricultural history is both well known and detailed, for in 1843 it became the site of a unique long-term experiment. It was planted with winter wheat, and so it has been every year since. In the early 1930s soil samples were removed to a depth of 6in, and were sieved and washed to concentrate any seeds present. The soil-seed concentrate was then bedded out being turned every 3 months and cultivated every 6 weeks. As the seedlings emerged, they were counted, identified and removed. From this study it was determined that this particular seed bank averaged 39,000 seeds in each square metre representing at least 47 species. A prodigious number in an environment in which regular disturbance and overwhelming competition with a pampered crop had been their pattern of existence for almost a century.

The seeds of weedy success but, as we have already seen, each one of these can be just the beginning of another story, for seeds may belong to annual, biennial or perennial plants. Although at first sight it might appear that to be a successful weed it might be best to be an annual, one has only to look out at the back garden to see that this is far from the truth.

Not all weedy habitats get dug up every year, the lawn being a prime example, and any lawnsperson will tell you that it is a never-ending fight to try to keep the greensward free of weeds. So turning to what must be some of the oldest 'lawns' in Britain, it is possible to see how different forms of management affect the make-up of the sward. In the valley of the River Thames near Oxford is a piece of grassland which has been managed in the same way for almost 900 years. Not far away are two other grassland plots which are on the same type of soil and are close enough to complete the perfect triplets, or at least the perfect triplicated experiment. The only difference is that the former has been managed as a pasture, grazed by cattle, horses and geese. (The record of their management is so good that we know that sheep have never selectively nibbled the greensward, and that during the years of the Civil War 1642–5 it was cropped for hay.) The other two have been regularly cut for hay in the late summer, after which cattle have been let in to glean and graze. The contemporary flora of all three fields was studied in 1937, and a total of 95 species of vascular plant recorded: 56 of these grew in the grazed pasture and of these 26 grew only there; 69 were recorded from the hay meadows of which 39 grew only within their confines.

Put yourself in the place of the plants and you will soon realise that under the grazing regime it would be a great advantage to keep your head down and especially to keep your growing points and dormant buds as low as possible. The pasture was indeed dominated by perennial grasses—Crested Dog's-tail, Perennial Rye-grass, Timothy and Rough Meadow-grass—with broad-leaved herbs like Dandelion and Bulbous Buttercup which form rosettes close to the ground, and others which spread by means of stolons and runners on or beneath the surface of the soil, like Self-heal and Creeping Buttercup. In contrast, in the hay meadows it doesn't really matter if you are tall and as far as competition is concerned the taller the better. Remember that you and all your companions are going to go for the chop at the same predetermined time, so that it will be best to flower and set fruit and seed before it happens. So it is that the hay meadows are dominated by annual grasses, especially the Bromes, mixed with erect leafy herbs like Ragged Robin and Yellow Rattle.

Perhaps the best and certainly the most extensive example of the effects of grazing was on the North and South Downs, which have for so long been the chosen locales for Sunday School outings and other recreational activities of the populace of Greater London. In the good old days before farming subsidies, the steep slopes and thin soils of the Downs were not considered worth putting to plough, and so were maintained in an open accessible state by the grazing of animals, especially sheep. With the demise of the lucrative wool trade and the increase of touristic activities, the sheep were removed, their share of the annual crop being taken by an increased population of rabbits. So the encroachment of shrubs and trees was held back and botanists and visitors alike were able to enjoy, as Charles Darwin himself had described, a floristic spectacle second to none. A profusion of orchids from the Early Purples and Frogs through to the Late Spiders and autumn Lady's Tresses all spring and summer long, bloomed in a community which may well have averaged more than thirty species of flowering plants to the

Primrose
Primula vulgaris Hudson

(plate 7)

square yard. Then along came the disease myxomatosis and almost wiped out the rabbit population. Immediately the shrubs and trees began to take over, moving in from woodlots, copses and hedgerows to suffocate and smother much of that diverse beauty. Indeed much of it would have disappeared if it had not been for the creation of a new breed of very special 'grazers and browsers'—the members of the British Trust of Conservation Volunteers who replaced the work of the jaws of both sheep and rabbits with scythes, sickles and elbow grease.

The impolite term used by the Trust's members to describe this activity, especially on a hot downland day, is 'scrub-bashing'; and anyone who has sweated to hold back the advance of modern forests will understand only too well the meaning of the term. They will also understand just how much work our early farmers put into clearing the way for the weeds, and how certain plants became adapted to everything which people and their animals both domesticated and introduced could sling their way.

Even if you are not an active member of the BTCV or your own Country Conservation Trust (and if you are not, shame on you), you can savour both these experiences right in your own back yard. Instead of border-to-border perfect fitted greensward, how about your own mowing meadow full of weeds (sorry, flowers). It will soon become a thing of great beauty, and management which consists of less use of the mower can't be bad, be it motorised or humanised. The golden rule is, only cut the grass once the flowers have set and ripened seed. You can of course speed the whole thing up by buying a packet or two of mixed meadow flower seeds, or by starting a seed bank of your own using local stock. That corner of your garden, or island in your lawn, will soon become much more than a blooming miracle, for it will attract a profusion of insects, birds and even bats.

Mind you, the mower does not compensate for all the effects of grazing and browsing animals, for whether it is a hover or a cylinder machine, it doesn't have hooves to dig in and hence cut up the turf. When grazing, an average cow walks 2–3 miles and an average sheep 3–8 miles a day, with hoof pressure of between 10.5 and 22.8lb/sq in. With four hooves at every stride, no wonder the turf and the weeds get all cut up. No wonder, too, that the majority of the perennial weeds and especially the grasses thrive on being regularly chopped up with a spade. If only Couch-grass hadn't been invented—no wonder its Latin name is *Agropyron*—and the same goes for Horsetails, Bishop's Weed and all the others which are overflowing into my rose bed! The fact is that they were not invented; they, or at least something very like them were already there, lurking members of natural plant communities. All they did was to seize upon each and every opportunity put on offer by farmers and gardeners alike.

There is, however, one rule which the weeds like all other plants have to obey, and it is best stated as follows. The more closely plants are crowded together, the smaller each plant is liable to be; the more plants are thinned, the bigger individuals can become, until they may ultimately reach the maximum size for the species. Gardeners will of course be saying, 'Teach your grandmother'; for this rule is the basis of the champion-plant syndrome which begets perfect crops of flowers and fruits, and that First Class Rosette at the Annual Show.

That this law can be expressed in strict mathematical terms which all plants, be they giant trees or small herbs, obey without more than mean deviation, may be of only academic interest; but when it comes to the growth of plants in natural communities and of weeds, it is of great practical importance. In nature there are no kind gardeners with their eyes on the local show to come along and thin the plants so that they grow to their optimum size. The same is true of weeds. So bearing all this in mind, to be a successful weed and to survive despite this law, plasticity must be a great advantage. A plant which is able to flower, fruit and bear seed whatever its ultimate stature must stand a better chance in the weed race.

To find one which appears to be perfect in every way, I don't have to look further than the flower-bed outside my window and page 823 of the *Flora of the British Isles*—its name is Groundsel. It is an annual, although it can overwinter producing an abundance of flowers next year. It has erect or ascending, weak, rather succulent stems which may be from $1\frac{1}{2}$ to 18in tall, and whatever the height it can both flower and set seed. Each fruit is ribbed, the ribs being very hairy, and has a long silk parachute. It can either be carried by the wind or, when wet, stuck onto socks or other sorts of fur or feather.

Common Groundsel (*Senecio vulgaris*)

In 1972 I did an experiment in my garden, in which I followed the progeny of the 78 plants of Groundsel I found growing in one square metre through the three generations they managed that year. If I had tended them all and continued planting their progeny out at the original density I would have ended up with 53cwt of viable flyable fruit. As each fruit weighs .0000070z that would make 13.5 American billion (13,500 million), a staggering thought, and without doubt Groundsel is so common because it is a weed for all reasons. It is, however, the following fact that clinches the matter and puts the Groundsel top of my garden weed list. The *Flora* states that it can flower in every month of the year; it is thus also a weed for all seasons.

(plate 8)

Yellow Loosestrife
Lysimachia vulgaris L.

Scarlet Pimpernel
Anagallis arvensis L.

PALAEOGRAPHY, PALYNOLOGY AND PALACE PREHISTORY

Prese-e-ent Arms! The crisp staccato of burnished boots on parade and the glint of gunmetal on polished wood snapped the whole crowd into expectant silence. Even the youngest, pressed against the railings, held their breath. Their wait had not been in vain; the Queen was coming. The silence of expectant homage and the warmth of the sun reflected from stone and tarmacadam transformed the roar of traffic in The Mall and Constitution Hill into a background buzz, like insects in a summer meadow.

Aaatishoo!! A mighty sneeze broke the silence. It had emanated from one of the Guards on parade, though not a bearskin nor even a hackle wavered to pinpoint its origin. It was, however, almost immediately forgotten, for at that moment a long black car caught all onlookers by surprise as it slid in through the gates, not out as everyone had expected, and the crowd roared its welcome. There was the Queen, and Prince Philip too, on their way home from where? At that moment it didn't really matter, there was the Queen of England, only feet away and waving straight at me—a right Royal second soon eclipsed by the shiny black boot of the car, strangely smeared with spots.

The human mind is a very peculiar thing, to be able to recall such trivial detail so many years after; for this happened in 1952. I can remember why I was in London. I had a ticket for the 222nd performance at the Royal Opera House, Covent Garden, of *The Sleeping Beauty*, and had decided to do the sights of London before going to the ballet; what is more, I got my first live look at the Queen—all red-letter events which should be remembered. But the sneeze and the spots on the car only came back in a flash as I was in the middle of planning this chapter. The date was there in my diary, 5 July, and the Queen had been out visiting. Both are facts of recorded history. That the sneeze and the spots were caused by products of the same species of plant is just surmise, but thereby hangs my tale.

Facts of history are, by definition, recorded either in writing or the printed word. They may therefore be suspect—for the recorder, diarist, translator, author, artist, compositor, scribe or plagiarist may have been dishonest, or a fool. Artefacts left by people or features contrived by nature on the other hand fall into the realms of pre- or natural history, and cannot in themselves tell lies, although the interpretation of their exact meaning depends upon the expertise of the researcher and the sum total of his or her knowledge at the time of study.

Important in our present context is that historical fact doesn't take us back very far beyond Domesday Book of 898 years ago which states that the area of the Palace Gardens was then the property of the Abbots of Westminster. Prior to this auspicious date, all we can do is to turn to the artefactual or natural evidence and try to find out more. It is fortunate for all students of prehistory that, long before William the Conqueror commissioned the writing of that great book, across his kingdom and far beyond nature herself had provided minutely detailed records.

To read and understand Domesday Book in its original form, it would help if you were an expert in palaeography. To read and understand nature's own history books you need great expertise in palynology. The former relates to the analysis of ancient writing, the latter to the analysis of pollen. It is my surmise that it was pollen which made the guardsman sneeze.

Anyone who, like he or me, suffers such allergy, will know to their chagrin that certain types of pollen are much more effective than others in bringing on the sneezes, weeps and runny noses, and that different people are affected by different pollen grains. We are, in fact, identifying the pollen, in our own sensitive way. Such differential allergy is due to chemical differences in both organisms, the pollen producer and the pollen receiver. Specific chemicals or complexes on the surface of the grain react with specific chemicals or complexes on your mucous membranes. Nothing more than interaction down at the level of the molecule— but enough to tickle your nasal fancy in a most explosive way.

Each and every pollen grain consists of a pack of chemical information, some lumped into the concept of genes, surrounded by two protective coats. The inner is made of cellulose which protects the living contents of every true plant cell, both from the environment and any would-be Vegans. Though a polymer of sugar molecules, cellulose is constructed in such a way that the living kingdom has been forced to many extremes in its attempt to get at the energy-rich food-stores held within. These range from the great grinding teeth of elephants and the long multi-stomached alimentary canals of ruminants to the efficient cellulose-splitting enzymes of bacteria and of certain snails. Facts to be ruminated upon by all humans and humanitarians. However, so important is the genetic message contained within the pollen grain that a second outer protection is provided by a layer of sporopollenin, a substance about which much less is known.

When a biochemist wants to find out about an organic substance he or she must break it down into its component parts, and then try to piece the chemical jigsaw together. The trouble with sporopollenin is that it is a very difficult molecule to crack open, rather like a jigsaw held together with a super glue which is stronger than the pieces themselves. To crack such a molecule the chemist must use tough measures—very strong oxidizing substances like sulphuric acid mixed with hydrogen peroxide or chromic acid. Oxidizing agents don't come much tougher than that; real 007 stuff unless the operator knows exactly what he's doing. So you adds your acid to your sporopollenin from behind a protective screen and goggles, you takes your chance and, if lucky, are left with a mixture of carotenoids— substances which include the pigment carotene which makes carrots the colour they are and helps us to see in dim light.

This rough, tough substance is not only found protecting the pollen grains of flower-bearing plants and conifers, but also performing similar functions round the spores of ferns, club-mosses, algae and fungi. In fact, throughout the whole vegetable kingdom it appears to be the proper pigment polymer for protecting plant propagules, hence its combination name, sporopollenin. Even more intriguing, sporopollenin-like substances have been found in some of the oldest rocks on earth dating back to 3.7 thousand million years ago, and thus older than the oldest known fossils. What is more, similar organic substances have been found in the Orgueil and Murray meteorites—not absolute proof, but good circumstantial evidence of life, or at least proto-propagule protectants, in other parts of the universe.

Although sporopollenin is so widespread, it is only in two groups—flower- and cone-bearers—that it is built into the complex-patterned outer walls which make their pollen grains such beautiful subjects for microscopical study, and make many of them as identifiable as the plants from which they came.

(left) Crested Dog's-tail (*Cynosurus cristatus*), (centre) Annual Meadow-grass (*Poa annua*), (right) Purple Moor-grass (*Molinia caerulea*)

Lesser Periwinkle
Vinca minor L.

(plate 9)

During their final phase of development within the pollen sacs or cones many other substances, aptly named the 'pollenkit', are laid down in the tough sculptured wall substances. These may include sticky and odoriferous materials which may help in the dissemination of the pollen by insects and other animals. They also include the 'recognition proteins' which may well have been the ones which made the guardsman sneeze, though they are there to perform a much more vital function for the plant. Their role is to ensure that the pollen grain only germinates with success on a receptive stigma of the right sort. If it lands on the wrong type of stigma then these specific chemicals will not be recognised and an 'allergic' reaction is set up, stopping or slowing the growth of the pollen tube so that fertilization cannot occur.

Such pollen recognition works in many different ways and has been of great importance throughout recent evolution. However, perhaps most important of all is the way in which in many cases self-pollination is avoided, ie the pollen from one flower cannot fertilize the ovule of that same flower. Such incestual tendencies could only lead to uniformity of type and less likelihood of change and so would have slowed evolution.

A review of the structure of flowers growing in any garden will reveal many varied mechanisms which help to stop self-pollination, showing that this was not the way ahead for the flowering plant. For example some plants, like Holly, produce male flowers on one individual and female flowers on others, thus causing many holly-bush owners much disappointment at the approach of the festive season. It can also cause problems amongst the plants themselves, 'frustration' and a great wastage of the genetic information in many potential embryos, but such is the importance of cross-pollination and this is the way of nature. A less extreme method is to have the two different sorts of flower on the same individual, so that each plant has the potential of bearing both fruits and seeds.

Then there are the complete or perfect flowers. In some of these the pollen sacs ripen and shed their contents while their stigmas are still immature and hence unreceptive to pollen. Such a mechanism is called protandry (male first) and is well illustrated by the flowers of Queen Anne's Lace. The opposite condition in which the stigmas are receptive and ovules ready for fertilization before the pollen is released is rarer, but is well demonstrated by the Ribwort Plantain. The flowers on its spike-like inflorescence open from the top down, the closed flowers first protruding their receptive stigmas which wither away before the long stamens start to show. This condition is known as protogyny (female first).

Some flowers have architectural features which prevent their pollen from being transferred to their own stigmas. The Iris is a good example, for in those fantastic flowers the ripe pollen sacs are sheltered beneath the branched petal-like style. Even the shy Primrose boldly demonstrates its heterostyly by which self-fertilization is minimised by the positioning of the style and stigma. The flowers all have both male and female parts but are of two different types, some being pin-, others thrum-eyed. In the former, which are best called long-styled flowers, the receptive stigma which looks like the head on a pin occupies the mouth of the flower tube hiding the pollen sacs which are set below. In the thrum-eyed or short-styled flowers, the arrangement is exactly opposite, the pollen sacs occupying the mouth of the tube with the stigma hidden below. The crafty bit comes when an insect visits the flower in search of nectar. Only one part of its body will rub against the stigma and another against the pollen sacs which, if ripe, plaster that part with pollen. On visiting the next flower pollen will only be transferred preferentially to the stigma if the flower is of the other type. Try it with a matchstick and see for yourself. Things may be even more complex in the Purple Loosestrife, where there are two circles of stamens and three possible lengths of style and hence levels of stigma.

Then there are the many mechanisms in which it would seem that both plant and insect have evolved together in order to overcome the difficulty. That flowers produce nectar which is rich in sugar and hence in energy and, what is more, go to the trouble and material expense of advertising the fact with coloured petals and perfumes, indicates that they must derive much benefit from the visitations of the

Purple Loosestrife (*Lythrum salicaria*)

pollinators. This is corroborated by the observation that plants which are pollinated by insects on the whole produce far less pollen per ovule that those which rely on the inconsistencies of wind pollination.

Taking some of our native plants found in the Queen's Garden as an example, Hazel produces 2,549, Beech 637, Lime 44 and Ribwort Plantain 15, pollen grains per ovule; the former two being wind, the latter pair insect, pollinated. The rule is not absolute, but good enough to be growing on with: the more you rely on wind the more pollen grains you should produce to ensure success; but if you expend a little energy in the production of nectar, pigments and aromatics, you can be more sparing when it comes to pollen production. There are, of course, other aids on both sides. A good wind-pollinated plant may not need to advertise, but long dangling anthers and feathery stigmas help to send the pollen down and trawl it from the wind. Likewise it is the wont of many wind-pollinated plants like the grasses to hold their flowers high above their leaves or, like so many of our deciduous trees, to flower before their leaves unfurl and crowd the flight paths.

The humble Nettle, not to be outdone, has its stamens bent rather like sets of little knees in pious genuflection. That is while in bud; at maturity they suddenly straighten, scattering pollen in all directions especially on a windy drying day. The doyenne of wind pollination must be the elongate catkins of the Hazel. Their scales collect and hold the pollen once it is shed from the anthers, only when the wind blows, wagging the catkins, are the pollen clouds released to be blown far and wide on the same wind—a much more efficient mechanism that just hanging out your stamens and hoping for the best. You can make it all happen for yourself by bringing catkins inside when young and letting them mature in peace, then wagging the branch and noting what happens.

One must be careful, however, not to jump to conclusions, for not all catkins are pollinated by wind. Take for instance those of the Pussy and other Willows; they are undoubted catkins, yet they all produce nectar and are visited by insects. In the same way, the foreign relatives of even our largest English friends the Oaks let us down when it comes to their mechanisms of pollen transfer. Both our British Oaks produce pendant catkins which open just before the first flush of leaves and are pollinated by wind. In contrast, their tropical cousins which grow as part of evergreen forests produce upright catkins which are pollinated by insects.

Everywhere you care to look there are things of great fascination waiting to be discovered about even the commonest plants which sustain and surround our lives. None more so than the ways in which our flowers and insects appear to have co-evolved to ensure that cross-pollination occurs. One example must suffice and it relates to a plant whose flavour is fine enough to grace the onions of any table. Its name, of course, is Sage, and here is the scenario of its pollination.

A large humble bee alights on the lower lip of the flower which is conveniently modified as a landing platform able to bear the weight of such a large visitor. The bee comes in search of nectar of which there is plenty held ready in the cup-like base of the flower. So far so good, and quite straightforward, but now we come to the complicated bit and so does the bee. Each flower has two stamens which are made of two very unequal halves, constructed as follows. One half is sterile and

Duke of Argyll's Tea-plant
Lycium halimifolium Miller

Bittersweet
Solanum dulcamara L.

(plate 10)

forms a short projection or fulcrum in the throat of the flower, this being joined by a long connective to the fertile pollen sac which is held up within and protected by the blue flower-hood. The long connective thus forms an unequal lever which sits astride and is movable upon the fulcrum. Now although the bee doesn't understand the theory of levers, when it pushes against the short arm its movements are magnified by the system, bringing the ripe pollen sacs slap bang down in the middle of its furry back. Try it for yourself with a pencil. The bee or HB then flies off to the next flower, which if it is in a more advanced state of development has its receptive stigma hanging down in exactly the right place to collect the pollen.

The first person to describe this truly novel mechanism was Konrad Sprengel in a book published in Germany in 1793. In this delightful work he not only described pollination of the Sage but revealed many other mutual relationships between the form and colour of flowers and the insects which frequent them. So fascinating were these revelations that the study of floral ecology, as it came to be known, bloomed out of all proportion. Yet despite the reams of descriptive data which have been added to the subject since that time, no one has ever answered the question as to which came first, the mechanisms or the pollinator? Chicken and egg? Well, not quite, and perhaps a look at the pollination of another common plant will help in our deliberation.

Heather in bloom, massed on the hills, is enough to take anyone's breath away, but how many of us ever bother to take a lens and investigate the beauty of each tiny bloom. Each one is a deep bowl some $\frac{1}{16}$th in across made of 4 pink sepals and 4 pink petals, the former being longer than the latter. These enclose 8 stamens (the pollen sacs of which are furnished with a projection or awn and a pore through which the pollen can escape) and a central style with a 4-lobed stigma. Copious nectar is secreted from 8 swellings which form an almost continuous ring around the base of the ovary. As the flower matures, the lower parts of each petal become succulent and as they thicken the blossom is forced open. The lower petals open more than the others and so the flower ends up with imperfect symmetry.

In order to reach the nectar, insects must be able to force their proboscises down between the filaments of the anthers which are also slightly swollen—just to make it more difficult, one is tempted to say. Even without taking such anthropomorphisms into account, it is a tight squeeze, and in consequence pollen is shaken out of the anthers. The awn-like appendages also get in the way, increasing the insect aggro and hence the pollen release. Nectar production is copious, and one can sense the excitement of the various insects as they forage out across the moors. Apiarists, butterflies, flies and many more get in on the act, but it is the former who have helped make these tiny flowers so famous, and given the world the real taste for heather honey.

Not all the insects are mere visitors, some like the heather-flower thrips are small enough to live out their lives within the shelter of the flower. They are indeed so small that they can creep down amongst the swollen anther filaments and sip the nectar. In so doing they not only get covered with the sticky substance, but they also wag the anthers and themselves get dusted with pollen. Now as far as the

(*plate 11*)

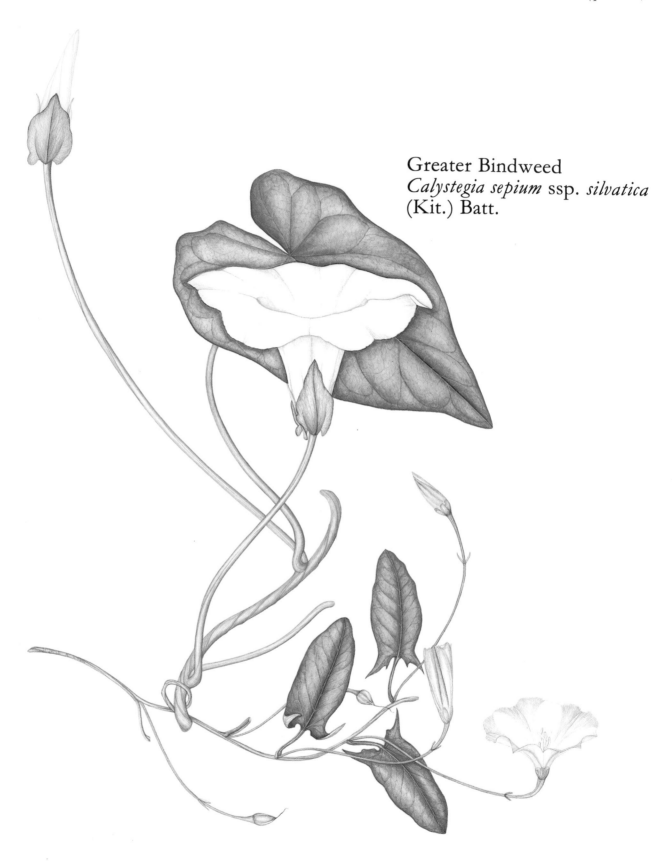

Greater Bindweed
Calystegia sepium ssp. *silvatica*
(Kit.) Batt.

Lesser Bindweed
Convolvulus arvensis L.

(plate 12)

Bellbine
Calystegia sepium ssp. *sepium* (L.) R.Br.

males are concerned, this is both hard luck for them and for the process of cross-pollination because, as they cannot fly, they stay put within one flower. However, the males are much rarer than the females and the latter have wings; so within any one flower there may be several frustrated females, but they can fly away despite the nectar and pollen load they are stuck with. What is more they use the stigma of the flowers both as a launch and as a landing pad and so are active both in self- and cross-pollination.

That is not the end of the story, for the female thrip, once happily married and mated, crawls back down into the base of the flower, sips more nectar and gnaws at the succulent flower parts to give her strength. Finally she lays four eggs in the base of the petals where they are protected along with the heather's own developing seeds by the persistent silver-grey flower bracts. So the life cycle of both flower and insect comes to a satisfactory completion.

Now comes the strangest bit of all, for despite the great diversity of insects which visit and use the flower, Heather is a prodigious producer of pollen. It may well hold the world record for it has been estimated that one square metre of good heather moor can produce 16,000 million pollen grains. That makes an average of 1,000 grains per ovule which, if all the thrips and bees and other insects did their work properly, would mean 16 million seeds from each square metre. However, they needn't really bother too much, for towards the end of its pollen production each flower produces less and less nectar, and as this happens the swollen filaments are able to elongate, allowing the pollen sacs to protrude from the protection of the flower. Hence wind can complete the job started by the insects—absolutely vital upon the mountains and moorlands where a cold windy summer may curtail the

Lesser Dodder (*Cuscuta epithymum*)
on Heather (*Calluna vulgaris*)

Hemlock Water Dropwort
Oenanthe crocata L.

Giant Hogweed
Heracleum mantegazzianum
Sommier & Levier
(detail of stem, fruits and seeds)

Golden Chervil
Chaerophyllum aureum L.
(fruits)

(plate 13)

activity of the insects. So it is that the Heather gets the best of all worlds, and we can perhaps see how neither the pollinator nor the mechanism came first; they both evolved together.

The insect simply looks for food, and the welfare of the plant and its future generations depends on effective pollination and especially on cross-fertilization. Variation in both the behaviour of the insect and the structure of the flower react together to give a better chance of survival to both their progenies. This is both the stuff and the mechanism of natural selection, despite the fact that the product may look so intricate that creative design would appear to be a more likely explanation. My only questions to those who still hold onto the hypothesis of special creation as it was preached pre-Wallace and Darwin, is this. Why limit your concept of a God to someone who must create each and every manifestation of life anew, and why if each is created with a purpose are there so many imperfections?

It is a lousy world if you are a lady thrip gravid with a new generation of quads, and you get eaten before those eggs can be laid within the protection of the Heather flower. It is a lousy world if you are the progeny of self-fertilization, and your development is marred by abnormality or monstrosity caused by genetic incest. It is a lousy world if you suffer from hay fever and must dread the hot days of summer when the grasses explode their pollen into the air. I cannot think that any God worth worshipping would sit back in the cold light of creation and design in such lousiness.

I can, however, both believe in and worship a God who set the process of creative evolution in motion, a process which through time will work out His purpose of perfection. Creation is a once and for all thing, and is therefore finite; creative evolution is on-going and hence infinite in its possibilities, the lousiness of imperfection being just a step along the way. I believe that we as part of that creative evolution have been set aside with the power of conscious, reasoned thought; that we alone can learn from history both natural and man-made and put that knowledge to good use. We have the power of God's creation in our hands and part of that creation lies in an understanding of the commonest flowers.

As you look at flowers for their adaptations, remember there are those which have no special mechanical devices nor timed development of their male and female parts to overcome the problems and limitations of self-pollination. They may however rely on the chemical methods of incompatibility such as those vested in their recognition proteins which have already been mentioned and which, because we cannot see them, have been much less well researched.

But to sum up their function, they either stop or slow the germination of the grain or the development and growth or penetration of the pollen tube, so that the genetic message of the male never reaches that of the female in time to complete fertilization. The ultimate is that the sets of genetic information inherent within the two germ cells are themselves imcompatible, so that even if pollination is completed, fertilization is impossible.

So we have within some of our simplest and most open-faced flowers, systems of communication so tiny that they make our most modern space-age electronic systems seem grotesque and obsolete. What is more, if that information gets to the

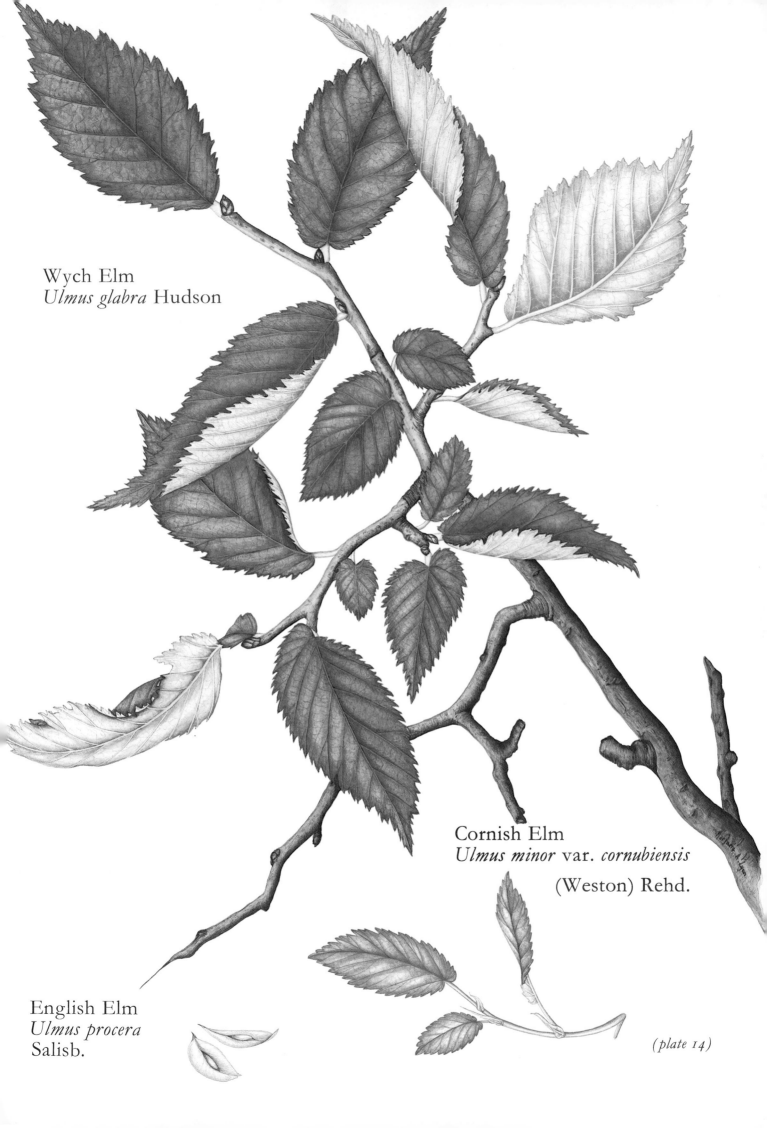

Wych Elm
Ulmus glabra Hudson

Cornish Elm
Ulmus minor var. *cornubiensis*

(Weston) Rehd.

English Elm
Ulmus procera
Salisb.

(plate 14)

wrong receiver it can cause distortion, even noise, even on a Royal occasion. Aaatishoo! That is exactly where we came in—well, not quite; for now we are armed with many facts concerning pollen which may allow us to understand the presence if not the meaning of all those natural pre-Domesday history books out in our countryside, waiting to be read.

The pollen of all plants is distinct and may be recognised both by the plants themselves and by the nasal membranes of certain people. Also, as mentioned earlier, thanks to sculpturing of the sporopollenin and the science of palynology, studies since 1793 have shown that many of the pollen grains can be recognised under the microscope. Thanks too to the resistance of sporopollenin to chemical attack and decay, pollen grains and spores rot very slowly and so may be found preserved in sub-fossil form in soils, silts, humus and peats. The amount and state of preservation depends on the ease of access of oxidising agents into the deposit upon which the grains fell; thus the best pollen records are found in deposits formed in waterlogged soils, ponds, marshes, swamps, fens or bogs in which stagnant water shuts out the oxygen-rich air. Here pollen is trapped year by year as sediment, silt or peat builds up, sealing in the record. Deposits formed in such places are eagerly sought by palynologists who, using all the sophistication of modern science and an awful lot of patience, open and date the pollen-record 'books' and read the ordered detail of their pages.

It is the reading of such detailed records found throughout Britain and far beyond which tell us that a mere 150,000 years ago much of the British Isles was laid waste by the penultimate advance of the glaciers of the last Ice Age, which had shaped the main course of evolution for almost a million years. No life survived below the ice which at that time reached southwards to a line joining the Thames and Severn. South of this tide of total destruction, the little that was left of the British Isles as we now know them was covered with open tundra vegetation rich in montane and arctic alpine plants. The climate was, however, on the mend, and as the glaciers melted plants and animals flooded back from the warmer south to take up residence in Belgravia and beyond.

The pollen records which give us some of this information were discovered during excavations in Trafalgar Square and its environs. Pollen of Hazel, Hornbeam, Oak, Pine, Elm, Birch, Alder, Maple, Spruce and Lime—in that order of abundance—were found along with that of many shrubs and herbs. The presence of the distinctive fruits of Water Chestnut and the distinctive pollen of the Maple (*Acer monspessalanum*) indicate a warmer climate than Londoners enjoy today. The richness of the vegetation of that warmer time is also borne out by the remains of elephant, hippopotamus and rhinoceros, all of which were found there. Big game living in Trafalgar Square, and only 100,000 years ago!

The presence of many such pollen records in different situations both above and below the contemporary level of the Thames, hinges on the fact that a huge sheet of ice weighs a tremendous amount. At the height, or was it depth, of the glaciation, at least 10 million million tonnes of solid water were pressing down on the land mass which is now England, Scotland and Wales. This depressed the land surface so that all-low-lying land should have disappeared beneath the surface of

Dogwood
Thelycrania sanguinea (L.) Fourr.

Elder
Sambucus nigra L.

(plate 15)

Lily of the Valley
Convallaria majalis L.

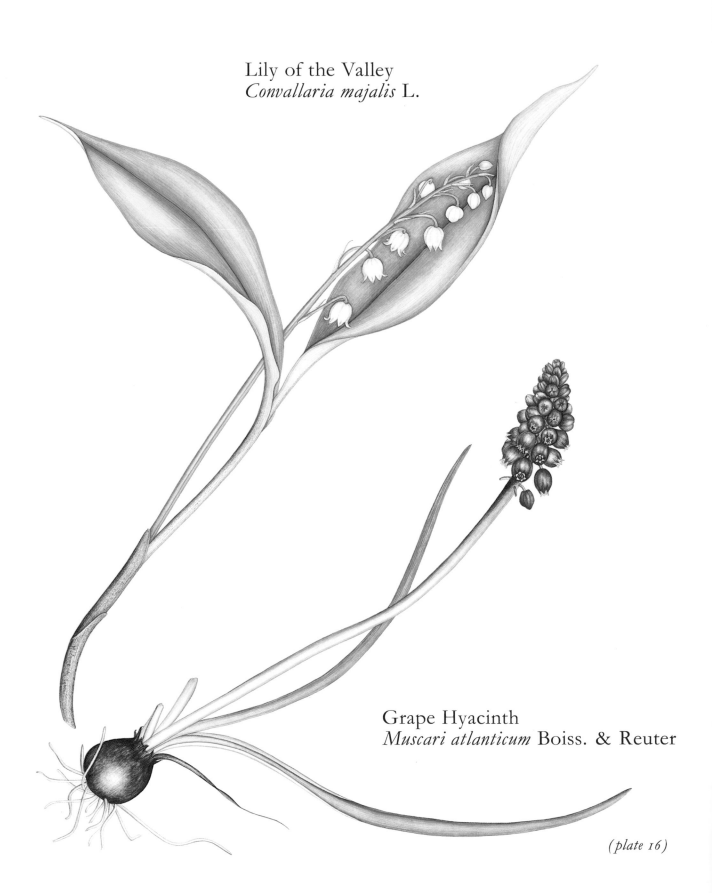

Grape Hyacinth
Muscari atlanticum Boiss. & Reuter

(plate 16)

the sea. However, as the ice originated as snow which itself originated by evaporation from the sea, the level of the latter also went down. The net result was that at the height of the glaciation a very differently shaped British Isles, with Ireland tacked on for luck, protruded as a peninsula north by west from Europe.

Thus, when the ice began to melt as the climate improved, it was easy for both plants and animals to move north by west into this barren land, taking up all the new opportunities on offer. Such free movement however only continued until enough melt-water had returned to the sea filling the English Channel and the Irish Sea once more. As the ice melted the burden was gradually lifted from the land which, heaving a great sigh of post-glacial relief, also returned towards its original position. Add to this the differential weight of the ice sheet from north to south and the consequent hinging of the land mass, and we get an ever-changing complex of levels, especially along the estuary of a river such as the Thames. These levels provided situations in which marshes and swamps could form, opening the pollen record for definite periods of time.

The Trafalgar Square and other pollen discoveries tell us that the warm inter-glacial period came to an end as new glaciers started to form in the mountains of the north, once more withdrawing water from the sea and pressing down upon the land. This new deterioration of climate was manifested in the pollen record by the rise in the abundance of both Birch and Pine, which eventually themselves disappeared, gradually replaced by the pollen of shrubs and herbs and other tundra plants.

This last advance of the ice was not as severe as the one before, the ice front reaching only just south of that hypothetical line which joins the Humber and the Severn estuaries leaving those areas now known as the Fens, East Anglia and the South East completely clear of ice. Yet it would appear from the pollen records that the presence of the ice in the North Midlands was sufficient to eradicate the vast majority of trees from the face of Britain, replacing the closed woodlands with open tundra.

Juniper, Dwarf Birch and Arctic Willow did their best to form a canopy, but in the main the London scene was dominated by Sedges, Grasses and arctic alpine plants including Alpine Poppy. Two-flowered Sandwort, the Snowy and Woolly Cinquefoils and the Arctic Buttercup, none of which now grows within the British Isles. What the vegetation lacked in stature it must have made up at least in part by its grazeability, for it supported herds of bison, reindeer, wild horse, mammoth and woolly rhinoceros. All these lived in England less than 50,000 years ago, and although we have little direct evidence, there is little doubt that such magnificent creatures made good use of what are now the gardens of Buckingham Palace.

Neither the build-up nor the demise of the ice sheets were simple smooth changes; there were both major and minor fluctuations all along the way, each one sending major or minor shivers through the vegetation and hence the pollen record. That the overall climate began to take a real turn for the better some 14,000 years ago is well documented through pollen, and we can say with complete conviction that somewhere between 10,300 and 10,200 years ago the tundra was in full retreat and warmth-demanding plants were once more growing in the

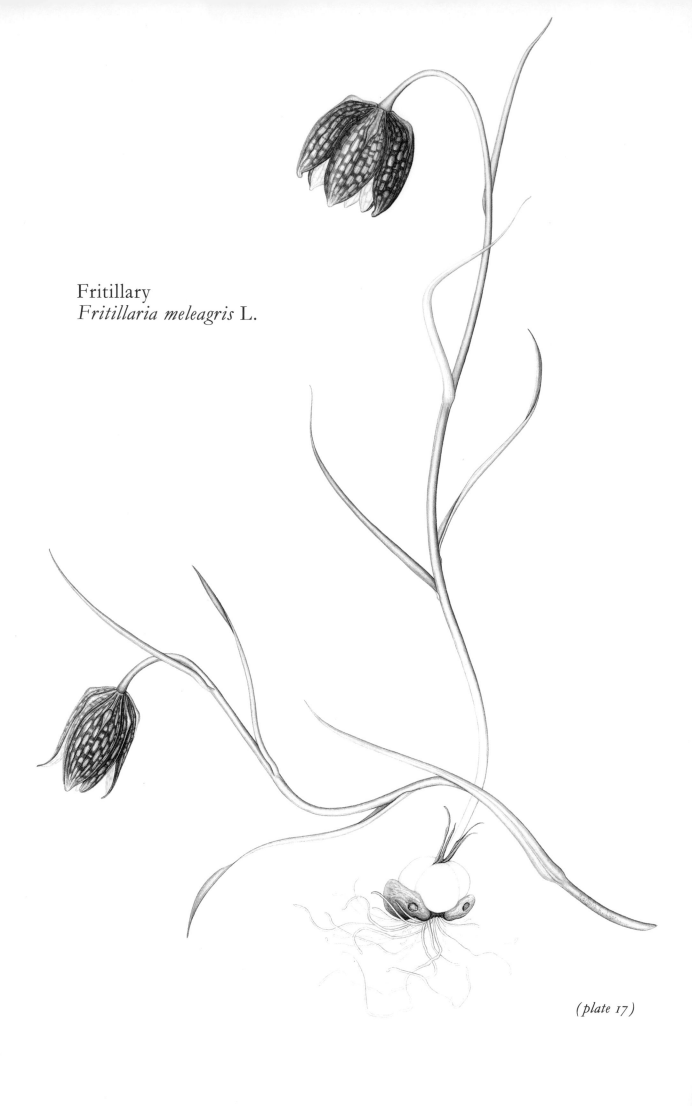

Fritillary
Fritillaria meleagris L.

(plate 17)

environs of Constitution Hill. Those on their way out included Docks, Mugworts, Thrift, Meadow Rue, Bellflowers, Knapweeds, Willow-herbs, Bedstraws, Plantains, Scabious and Valerian. The woody plants which took their place in order of abundance were Dwarf Birches, Sea Buckthorn, Juniper, Downy Birch, Silver Birch, Willows, Hazel, Pine, Elm, Oaks, Alder, Lime and Hornbeam. They migrated in from Europe across a land bridge which existed right up until about 5500 BC when the melting ice once more refilled the Channel, and the British Isles almost as we know them now came into existence.

During this period when the improving weather allowed more and more warmth-demanding species to grow, it is interesting to compare the order in which the main trees returned to their present limits of distribution within Britain as evidenced today. Juniper came first and is today found growing up to 3,200ft above present sea level; Silver Birch came next (2,500ft); Willows next (2,000–3,000ft); Hazel (2,000ft); Pine (2,200ft); Elm (1,000ft); Alder (1,500ft), its distribution is part controlled by water which itself can act as a heat store. Lime and Hornbeam have no given altitudinal limit, being confined to the lowlands. The former grows naturally as far north as Yorkshire, the latter only making it into Oxfordshire.

It is also of interest to note that, on average, a rise of 300ft brings about a drop of 1 degree C (2 degrees F) in average temperature. So it would seem that an overall drop in temperature of a mere 10 degrees C (18 degrees F) below that of today would banish all the trees from Britain. In comparison, a rise of a mere 5 degrees C (9 degrees F) above today should be sufficient to cover the whole area in broad-broad-leaved deciduous woodland. The pollen records tell us that is exactly what happened in the past, for 7,500 years ago, at a time which has come to be called the Climatic Optimum, Elm was a common component of the forest which then covered the highest peaks of the Pennines around 1,800ft above its contemporary limit.

So what of the forest which covered the Queen's Garden during that time of Climatic Optimum, which lasted from about 5500 BC until about 3000 BC? The nearest pollen records from which we have relevant sequential data are from beneath New Palace Yard, Hampstead Heath, Epping Forest and Crossness down the Thames, and though the story told by each is slightly different, they all give us a picture of a similar dominant type of forest. In terms of the absolute number of pollen grains the record tells us of the presence of the following trees: Oak, Hazel, Lime, Elm, Alder, Pine, Birch, Hornbeam and Willow in that order of abundance, with very few grains of herbs and other plants of open ground.

However in the knowledge that the trees which are pollinated by insects produce far less pollen than those which rely on wind, what is the true picture of the makeup of the forest? Of all the trees mentioned above, Lime and Willow stand out like a nectar-sweet thumb, for they alone are pollinated by insects. Lime produces only 44 pollen grains per ovule and, what is more does not like to grow too close to water. It also flowers in summer, early July to be exact, when the weather is warmer, when pollen decay would be more rapid, and after all the others have released their wind-borne pollen. So even if Lime had been the dominant tree

(plate 18)

Bluebell
Endymion nonscriptus (L.)
Garcke

in these forests, its pollen would have been under-represented especially if the pollen records were formed under waterlogged conditions, as they were at all the sites. In the same way Hazel, and especially Alder which thrives in and around such wet spots, would be over-represented and, what is more, would form a barrier filtering off the pollen produced by the trees growing on a higher drier ground before it reached the wet spot.

To appreciate the barrier aspect all you need do is walk into a copse or woodlot on a windy day. From the turbulence of the air outside, which tends to carry your hat let alone the pollen away, you enter a world of relative calm in which little eddies creep, reverberating within the shelter of trunk space. Trunk space is a scientific term denoting the volume of space contained within and protected by the tree trunks and their high canopy. I cannot however get away from the feeling that it is a much more down to earth homely thing, for a woodland is just that—a high-rise home for an enormous diversity of plants and animals, and a protected place within which the movement of even the lightest pollen grains will be contained. This is of little consequence to the trees themselves for enough pollen will be carried far enough to bring about cross-pollination. However, as far as the palynologist is concerned, the presence of trunks, branches, twigs and leaves within the trunk space actively inhibits and blocks much long-distance transport, making the interpretation of any pollen record more difficult.

For a long time the vanguard of palynology did not recognise these home truths, and when deposits were found containing abundant grains of Lime, they sought explanations in local variations of forest make-up, special soil conditions or even differential pollen preservation. To overcome these now evident limitations, much research has been carried out concerning the amount of pollen produced by each type of tree. English Oak came top of the list on a per-flower basis with an incredible 1.2 million pollen grains—blooming marvellous. In terms of pollen produced per tree over a 50-year period, Alder came top with around 4.3×10^{11}. As for Lime, it did very well, not far behind Alder with 3.3×10^{11}; and so it was suggested that in order to obtain a correct appraisal of the status of Lime in any forest as determined from the pollen record, the absolute number of grains must be multiplied by 8 in comparison with Oak.

Thus we find for the four pollen records in closest proximity to Buckingham Palace and studied to date, the corrected importance of Lime pollen is 58 per cent for Crossness, 80 per cent for New Palace Yard, 81 per cent for Hampstead and 91 per cent for Epping. This is evidence enough to suggest that Lime was a dominant tree in the forests of London during the time of the Climatic Optimum. Indeed, when the whole British pollen record is reviewed in this new limelight, a pattern of great tracts of Lime-dominated, or at least co-dominated, forest emerges. These Lime forests were centred upon the Midlands around Gloucestershire, where fine examples exist to this day especially on the banks of the River Wye. Even at the time of the Climatic Optimum which favoured such warmth-demanding species, the pollen records all show that Lime was limited to low altitudes. Another major area of Lime forest was in Lincolnshire, where it formed a broad wedge between the wet Fens to the south and the wetlands of the Humber estuary to the north.

57

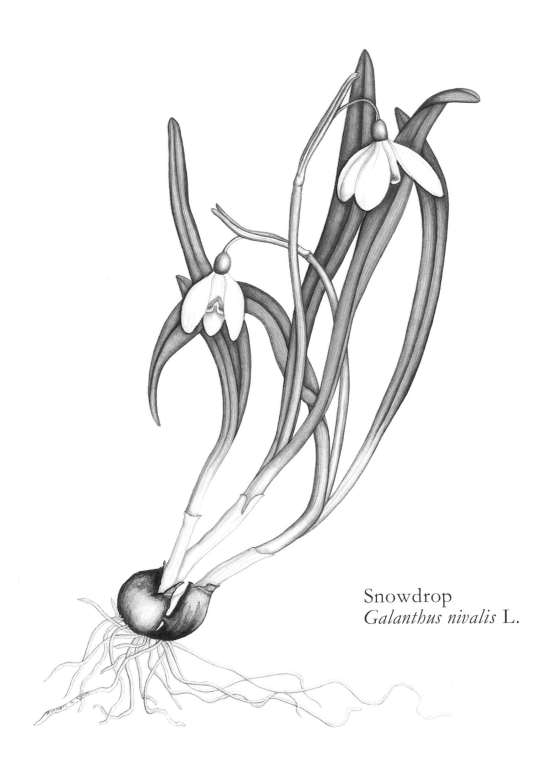

Snowdrop
Galanthus nivalis L.

(plate 19)

The third main area of Lime domination centred on the South East with the highest frequencies from areas underlain by chalk, or lime-enriched strata like the Bagshot sands. The greatest area of contemporary Lime woodlands is in Suffolk, where stands still actively coppiced are among our most beautiful examples of man-managed vegetation.

The original wide distribution of Lime woodland is upheld by the much rarer findings of Lime fruits, leaves and wood and remains of a small beetle which feeds only on Lime. It also comes from place-names which have their derivation from ancient words which themselves mean 'lime' or 'lime woodland'.

The origins of place-names depend on painstaking research into ancient documents to discover the oldest mentioned name, and on knowledge of the ways in which words have changed over the years. Such research has shown that the main groups of names revolve round the words, 'lind', 'linden' and 'bast', the latter being the name given to the strong phloem fibres which once were harvested from the trunks. Care must however be taken to avoid confusion with names derived from *lin* meaning 'flax' in Old English, *llyn* ('water' in Welsh) and *hlinc* ('hill' in Old English). However when the experts have taken everything into consideration the accepted limy place-names group well with the pollen records of the past, so here the disciplines of palaeography and palynology stand side-by-side in evidence.

To complete the picture we must now turn to the local conditions of the Royal Garden itself. The vast majority of its 48.59 acres is man-made and has been man-managed for at least 300 years. The whole is situated on a flood plain of the River Thames, which consists of river gravels underlain by London Clay. The latter which is both calcareous and retains moisture, approaches the surface as the garden slopes up towards Hyde Park Corner. There is thus no reason to doubt that the whole area was once covered with mixed deciduous woodland within the make-up of which Lime was in abundance especially in the drier parts.

In order to discover what other plants may have been present in the area at that time, I decided to list the plants found growing in abundance in contemporary Lime woods growing on similar soil in areas of the Continent which enjoy a climate similar to that which pertained in southern England during the Climatic Optimum. They are as follows: L signifying that they have been seen growing in the Palace Gardens, and P that pollen of that species, (P) pollen of that genus, or (P̲) pollen of that family, have been found in the London pollen records so far studied:

Alder	L	P	Cleavers	L	(P)
Baneberry		(P)	Cock's-foot	L	(P̲)
Barren Strawberry		(P̲)	Dame's Violet		
Beech	L	P	Dewberry		
Bird Cherry		(P)	Dog's Mercury	L	
Bishop's Weed		(P)	Durmast Oak	L	(P)
Blackberry	L	(P)	Dutch Rush		
Black Poplar			Early Purple Orchid		
Carex brizoides		(P)	Elder	L	P

Enchanter's Nightshade	L		Pignut	L	(P̲)
Field Rose	L		Prickly Shield Fern		(P̲)
Finger Sedge		(P̲)	Queen Anne's Lace	L	(P̲)
Gagea spathacea			Ramsons		
Galium sylvaticum		(P)	Red Campion		(P)
Giant Fescue		(P)	Redcurrant		
Goldilocks		(P̲)	Red-veined Dock		(P)
Gooseberry			Remote-flowered Sedge		(P̲)
Great Butterfly Orchid			Sessile Oak	L	(P)
Great Stitchwort		(P̲)	Small Balsam		
Guelder Rose			Small-leaved Lime		(P)
Hawthorn	L		Snowdrop	L	
Hazel	L	P	Solomon's Seal		
Hairy Bitter-cress	L		Spiked Rampion		
Hairy Woodrush			Stinging Nettle	L	P
Herb Paris			Sweet Woodruff		(P̲)
Herb Robert			Touch-me-not		
Hogweed		(P̲)	Townhall Clock		
Hornbeam	L	P	Tufted Hair-grass		(P̲)
Knotted Figwort	L	(P̲)	Twayblade		
Lady Fern		(P̲)	White Helleborine		
Lathyrus vernus		(P̲)	Wood Forget-me-not		
Lesser Periwinkle	L		Wood Horsetail		(P)
Lords and Ladies	L		Wood Melick		(P̲)
Male Fern	L	(P̲)	Wood Millet		(P)
Marsh Hawk's-beard		(P̲)	Wood Sanicle		
Mezereon			Wood Sedge		(P̲)
Narrow-leaved Bitter-cress			Wood Sorrel	L	
Narrow-leaved Helleborine			Wood Spurge		
Oak Fern		(P̲)	Yellow Archangel		
Pale Wood Violet			Yew	L	P
Phyteuma nigrum					

Angle Shades (*Phlogophora meticulosa*) caterpillar and
moth on Common Ragwort (*Senecio jacobaea*)

The presence of so many of these plants either growing in the Palace Gardens today or represented in the pollen record does not of course prove that their contemporary populations are the direct descendants of the originals—far from it. In fact it is safe to say that none are of such ancient stock. In like manner neither can we say that all those species of plants once grew within the bounds of what is now the Palace Gardens, although there is much less reason to doubt the latter for all the species have been recorded in the past within the London area. All, that is, except the five whose names are given in Latin. The reason for this is not to hide the facts, but simply that as they are not native British plants they do not have native English names.

Whether they ever grew in post-glacial Britain we cannot say for certain; but as we have no evidence of their presence then, they cannot be called either native or naturalised—a distinction which is of much more than academic importance. As the climate warmed and our native flora and fauna flooded in, the seas gradually returned to claim their own and eventually, as we have said, around 5500 BC Britain became an island once more and our total of native plants and animals became more or less fixed from that time. Native plants are by definition those which arrived on the scene under their own steam, which in this case means migration across the land bridge. Once the marine barrier was in position, land plants and animals had to be helped across in sufficient numbers to ensure that they could gain a foothold in competition with the flora and fauna already there. This was no easy job for, as we have seen, by the time the land route was closed most of the British Isles was covered with mixed broad-leaved forest. Despite all the fruits and seeds which sport dispersal mechanisms and the almost endless stream of migrant birds which crowd our air routes, the vast majority of these later arrivals which eventually made their home in Britain were introduced by one animal, and that wasn't able to fly the same route until 1909.

There is evidence that people first came to Britain about 300,000 years ago to enjoy the warm productive benefits of the last interglacial period. They were also back here hunting in the ice-free tundrascapes of the south during the last period of glacial advance. With the melting of the ice they came in increasing numbers, wet-footing it across the ever-diminishing land bridge from the continent. These were our direct ancestors—'yer actual indigenous native Ancient Brit'—who soon made themselves at home across the game-rich landscape clothed in the skins of animals and perhaps the products of a plant called Woad. At first they lived close to the coast moving up rivers and out into the woodlands, hunting, gathering and fishing as they went. At that time they were so much a part of the natural system that their presence is hardly recorded even by the best of the pollen books.

By the year 3000 BC a new human culture best called Neolithic had arrived upon the scene by boat and was well established especially in Ireland. It is now known that these Neolithic people were farmers and that part of their lifeway depended on the husbanding of cattle. These were fed, at least at first, with what else in a totally forested landscape, but the leaves of trees.

So it would seem that about this time pollarded trees began to make their appearance on the British landscape, for the best method to ensure a good crop of

(plate 20)

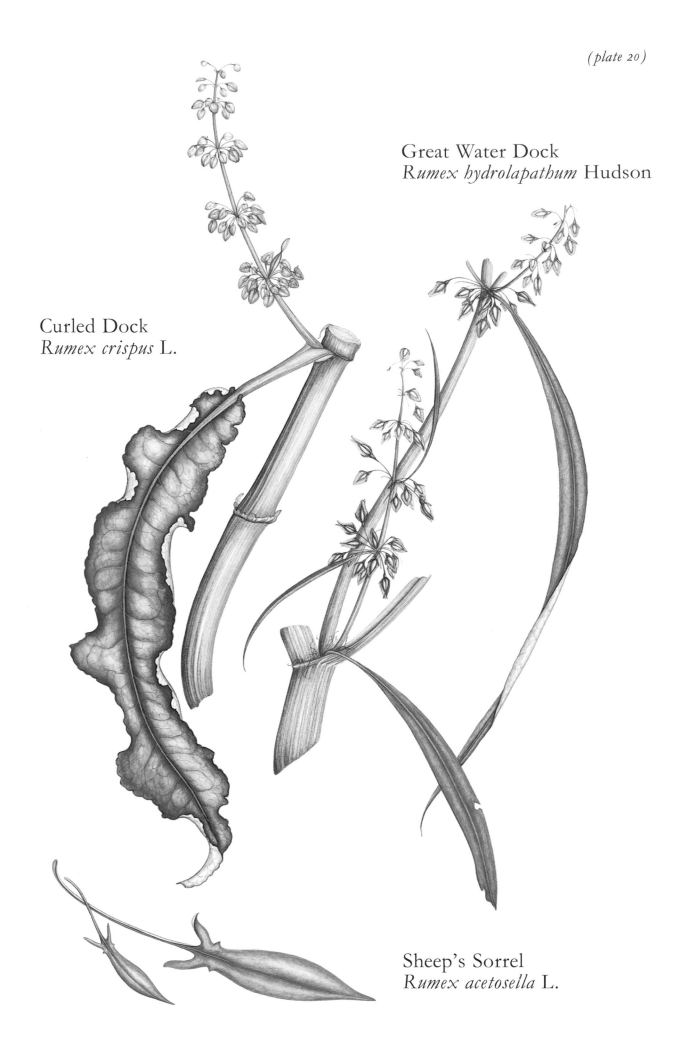

Great Water Dock
Rumex hydrolapathum Hudson

Curled Dock
Rumex crispus L.

Sheep's Sorrel
Rumex acetosella L.

readily harvestable leafy shoots each year is to prune the branches just above cattle-browsing level. This is exactly what pollarding is and the crop of leafy twigs so produced can either be fed to the cattle directly or stored for winter feed.

Pollarding is not the easiest job to accomplish especially with nothing more sophisticated than stone implements, and so it seems reasonable to assume that the farmers would have chosen trees which responded well to such treatment and produced a good worthwhile return for their effort. Such practice would of course drastically reduce the production of flowers and hence pollen by the trees in question, and so should show up in the pollen records.

The pollen records across the realm are unanimous that at first Elm was the chosen tree. And it would have been the best choice for not only does it respond well to such treatment, but of all our native trees it produces the heaviest crop of most nutritious leaves rich in both protein and carbohydrate, one mature Elm tree producing leaves equivalent to 2 acres of ungrazed meadow. What is more, pollen remains found in the close vicinity of Neolithic huts and pollen in material adhering to polished stone axes and other implements all show good correlation, and using the most sophisticated techniques provide a variety of dates, but all around the same period. All in all they show unequivocally that as the polished stone technology expanded, so the Elm pollen declined.

There could, of course, be other explanations. Acidification and improverishment of the soil from 10,000 years of exposure since it was uncovered by the melting ice could account for the phenomenon, for we know that Elm is rather fond of growing on base-enriched non-acid soils. However, as in most places the Elm was replaced by Ash and Beech, both of which exhibit similar soil preference, this explanation seems unlikely. Overall climatic change could have been another factor, but surely this would have been gradual, starting in the north and slowly affecting the forests further south. The pollen books indicate no such ordered sequence. Dutch Elm disease or some other epidemic of vast proportions certainly could have caused such a non-phased pockmarking of the Elm pollen record. With our own contemporary experience and as we have no real proof either way, this must remain as a possible explanation. The overwhelming evidence, however, points to Neolithic culture hacking away at the trees. This conclusion is backed up by the appearance of other microscopic evidence in the records of that time. As the Elm pollen declines there is a distinct rise in the pollen of Herbs, Grasses and Sedges and the spores of Bracken and other ferns—exactly what would be expected whatever opened up the tree canopy and let the light in. However, along with this new non-arboreal pollen comes that of Nettle and Ribwort Plantain, two plants which undoubtedly enjoy and to some extent depend on the company of man. This is especially true of the latter, for its other common name is Way Bread, and in North America where it was taken and spread by the settlers it is known to the Red Indians as Englishman's Foot. So here we have it, the first real record of real weeds.

You can imagine the excitement of all the palynologists and archaeologists at such wide-scale synchronised meaningful discoveries. More and more university departments and research institutes invested in peat borers and all the other paraphernalia of pollen analysis; the best, if not the biggest, departments, adding the sophistication of atomic age technology in the form of radio-carbon techniques to help them date their finds. In consequence more and more exciting evidence came to light and, as it did, other effects of human presence became evident. Some of the pollen records showed that along with Elm, the pollen of all the trees declined, their place taken by Herbs and Grasses, including the cultural (weedy) element. However the phases of open vegetation did not last long and were soon replaced by mixed forest once again; and painstaking work showed that such opening up was often of very limited extent, only a few acres or less being cleared at one time. The most likely explanation of such local hiccups in the pollen picture was put down to the practice of primitive nomadic, slash and burn agriculture; the local farmer clearing the forest, burning the brash and growing crops for a few years until the soil was improverished, and then moving on to start the whole cycle again.

Much argument ensued as to whether it was possible for people with nothing more sophisticated than polished stone tools to clear fell even small areas of natural woodland. The arguments revolved around the idea that pollarding a few trees was all right, but felling mature hardwoods like Elm and Oak was impossible. To stem the tide of argument, a group of Dutch palynologists equipped themselves with real polished stone axes from the museum in Copenhagen and sallied forth into the forest. They found that three men could clear almost 718sq yd of forest in 4 hours. They went further with their experiment, raising a crop of primitive Emmer Wheat in the ash-enriched soils. What is more, they did so well that they raised an economic crop—a feat which they could not repeat three years later when the local climate had done its worst to the bare soil. And as if to pat them on the back and reward their Trojan efforts, many Neolithic pollen records have revealed the presence of undoubted grains of cereal and others which look remarkably like those of Parsnip.

At last we are getting somewhere—real weeds, real crops, real kitchen-garden plants appearing on the local scene in sufficient quantity to be taken down by the 'local pollen blower' to be used in future evidence. People and their weeds were here to stay.

Meanwhile, back to those pollen records which give such evidence concerning the environs of London. The best pollen sequences studied to date come from a swamp on the edge of West Spa on Hampstead Heath and from several small valley bogs in Epping Forest. All show the presence of mixed forest with Lime as an important and probably dominant component, along with Oak and Hazel. They also record a reduction in the abundance of Lime pollen along with a marked increase in the amount of species more typical of open agricultural and ruderal vegetation. The presence of charcoal in some sequences indicates the use of fire; and an increase in the rate of deposition of sediment in others shows that erosion from the bare soil had been increased. Impoverishment and acidification of some of the soils also dates from the time of these Lime clearances, as Hazel, Birch, Oak and eventually Beech and Hornbeam, took its place during the regeneration phase. As all these clearances were localised phenomena, and as we have no such sequence from the Palace Gardens themselves, the dating of these first London Lime clearances are of little direct consequence. We can, however, say with conviction that some time between the late Neolithic (circa 2000 BC) and the Anglo Saxon period (circa AD 600) the forest which covered what is now the Queen's Garden was opened up for animal husbandry and then for agriculture.

These dates conveniently span the period up to the Middle Saxon era which may be regarded as a time of major settlement and cultural change in the area, analogous to the settlement of North America in the eighteenth and nineteenth centuries. This was also the period which saw the origins of the manorial system of which, as already said, we have written evidence.

Birch (*Betula pubescens*)

However, before we turn to history we must take our final look at the records in the pollen sequences and see what they tell us of the weeds. The evidence is both clear and detailed, for as the trees were first pollarded and then felled, the following plants came onto the London scene and gradually increased in importance: Beech, Elder, Holly, Hornbeam, Ivy, Mistletoe, Ribwort Plantain and White Water-lily—we are sure of the presence of these species. The next group includes all those which have been identified but only to generic level, and so these records may be made up of one or more species: Bur-reed, Dock, Maple, Mugwort, Nettle, Plantain, Reedmace and Willow. Then there is a large group whose pollen grains have been identified, or indeed are identifiable, only down to the level of the family, so that each record may well represent several genera and many species: Buttercup, Campion, Daisy, Dock, Fern, Goosefoot, Grass, Heather, Parsley, Pea, Primrose, Rose and Sedge. They came, they seized every opportunity open to them and they conquered, living it up to the space opened up by man for agriculture, horticulture or just for leisure.

The record of their success does not end here, for many smaller pollen records have been found at other locations much closer to Buckingham Palace. Unfortunately none of these provides a sequence of changing events, but they do give us an insight into an ever-changing London scene. From Wilson's Wharf in Southwark we have a glimpse of the vegetation as it was in the late Bronze Age, that is between 2570 and 3010 years BP. There the vegetation was open, rich fen or water meadow and adds evidence of the following plants: Iris, Kingcup, Meadowsweet, Reedmace and Yellow Loosestrife; with Bracken, Cereals, Docks, Goosefoot and Knotgrass growing above flood level.

We know that Julius Caesar himself recorded that the locals of south-east England in the first century BC lived in a woodland environment. Yet excavations from the Temple of Mithras site in the City of London indicate open vegetation and add the following to the plant list: (Anemone), (Bedstraw), (Bird's-foot Trefoil), (Bur-marigold), (Chamomile), (Clover), (Corn Spurrey), (Dandelion), (Dogwood), (Field Pansy), (Hardheads), Heather, Horsetail, Long-leaved Plantain, Male Fern, (Mallow), Marsh Pennywort, (Meadow Vetchling), (Milk Vetch), (Mint), (Pink), Polypody, Purple Loosestrife, (Rest Harrow), Scabious, Stag's-horn Plantain, (Stonecrop), (Vetch), (Whitebeam), (White Campion), (White Dead-nettle), (White Mustard) and (Yellow Rattle). Brackets in the above and in subsequent lists indicate that identification is only down to pollen type rather than an exact generic or specific identification.

The conclusion of the report on this site states:

> The bulk of the pollen was derived from three principal environments, arable cultivation and associated weeds, weeds from open and waste ground, and a marginal aquatic fen component. With the exception of Walnut, no introduced 'exotic' or garden plants are indicated from the pollen identified.

The presence of Walnut pollen from within the strata of the Temple, dating to circa first century BC, is of great interest for it has long been considered as an introduction from the Mediterranean during Roman times.

The final sites are the nearest to the Palace, being from New Palace Yard and dating from a little before Domesday Book (1086), and Broad Sanctuary dating from the fifteenth and sixteenth centuries. It is fascinating to think that those pollen grains could contain molecules of carbon which were last breathed out by William the Conqueror, Henry VIII or any of his wives. Records from both sites show that much of the dry ground had by then been cleared and turned over to agriculture or urban use. Their lists of weeds are impressive and add the following to those already accrued: (Aster), (Bellbine), (Bittersweet), Black Bindweed, (Bramble), (Bush Vetch), (Cherry), (Chickweed), Cornflower, Greater Celandine, Great Reedmace, Knotgrass, (Lesser Reedmace), (Lesser Salad Burnet), Persicaria, Poppy, (Red Bartsia), (Sanfoin), Scarlet Pimpernel, (Self-heal), Stinging Nettle, (Stitchwort), (Thistle), (Toadflax), Wood Vetch, (Woundwort).

Dorset Heath
Erica ciliaris L.

Cross-leaved Heath
Erica tetralix L.

Heather
Calluna vulgaris (L.) Hull

(plate 21)

Three other plants are included in these final lists and they are all definite introductions. The first is Flax, in all probability of the variety which was cultivated for the production of linen and linseed—the former to bed down ladies and gentlemen, the latter to feed horses and extract oil. The presence of Hemp, perhaps better known as Marijuana, doesn't prove that local society members were going to pot; rather that they were using it to make ropes to hang themselves in other ways. The third was Buckwheat, another introduction from the Continent, which will grow in the poorest, most acid ground. It was used to make buck-wheat flour—a poor substitute for the real thing, but useful in times of hardship.

Common Fumitory (*Fumaria officinalis*)

Whether all these plants were growing in or near the locality at the time is difficult to say. They could have been imported in fodder, thatch or other building or packing material. However we can conclude that by the time of the first printed record the vast majority of the plants which now grow in the Queen's Garden were already well-established in the wastelands and on the trade-routes to London.

The pollen of Lime was also present at both sites and the trees must have been quite common around New Palace Yard as they still are today in the Palace Gardens and other parts of London. That is why it is my surmise that Lime pollen made the unidentified guardsman sneeze. The spots on the boot of the car could well have been the fault of the chauffeur, that is if he parked or drove under a Lime tree en route to or from the Palace. Mind you, he would only be partly to blame for though Lime trees produce the sticky substance, it is the avid attention of many thousands of aphids which sends it on its way to the ground. Aphids live on plant sap, which is a very dilute solution of sugar. In order to obtain sufficient energy, they have to suck a lot of sap and what they don't use they excrete as honey dew. Drops of this sweet sticky substance then hang from their tails eventually falling, in this case to complete the second part of my story.

A GARDEN STEEPED IN HISTORY

Even in these 'enlightened' days of mechanical rotovators and selective herbicides, it is impossible to have a garden without weeds. As it is equally impossible to have garden weeds without gardens, it goes without saying that within the history of one lies the 'natural' history of the other. What then do we know of the exact history of the plot of land which today comprises the Queen's Garden?

The land upon which the Palace and its Gardens now stand was once part of the Manor of Eia and was, as mentioned in Domesday Book, the property of the Abbots of Westminster. The manorial system of land ownership and tenure was formalised in the reign of William the Conqueror (1066–87), who divided the country into portions of land—manors—which he gave to his barons and abbots. They owned and ruled these manors which were partitioned into an area of land—usually walled and fortified—containing the manor house, and another area—fenced or ditched—which was worked by the common people or villeins. Here they grew crops in rotation, usually wheat, then barley or beans and then fallow, the latter being full of weeds. For the service so rendered, they were provided with protection (for it was an unlawful country) and rights over strips of land which they could call their own.

Knowledge of agricultural and horticultural methods was fostered by men of learning who were then usually of the Church. They were the only people who could read the agricultural texts of Cato, Varro and Virgil which had been passed down in written form over the centuries, and were housed in monastic libraries. It is therefore probably safe to conclude that the first real gardens developed within the sanctuary of the walled monasteries. There, flanking a simple fishpond which provided both food and recreation, there would be an orchard and possibly a vineyard. Though a few flowers were cultivated, especially roses for use in religious ceremonies, the main splash of colour was to be found in the vegetable garden and particularly in the herbarium where roses, violets, poppies, lilies and the like were grown for medicinal use. These were all very small areas. Beyond the walls, fences and ditches, the common land on which animals grazed merged into wilder areas and wildwood, which was hunted for a variety of game and from which fruits, nuts and wood for fuel and other uses were gathered, in due season.

One development which began to formalise the landscape still further was the institution of the deer park; the classical meaning of 'park', both in England and

(plate 22)

Rosebay Willow-herb
Chamaenerion angustifolium (L.) Scop.

Pale Willow-herb
Epilobium roseum Schreber

Great Willow-herb
Epilobium hirsutum L.

(plate 23)

Common Knotgrass
Polygonum aviculare L.

Black Bindweed
Bilderdykia convolvulus (L.) Dum.

Wales, being an enclosure for semi-wild animals. Such a park was a rich man's privilege and a valuable asset, for it kept him and his family in fresh meat throughout the winter. The problem was that deer, even the small fallow kind which had been introduced by the Romans, are as strong as pigs and as agile as goats. Any park boundary, therefore, had to be sound; and it was usually a deep ditch backed by a mound topped with a hedge or a fence of cleft sticks. Such boundaries took an awful lot of maintenance, one reason why the majority of such parks have rounded not square outlines, thus minimising the work. Most parks contained, at least at first, various areas of woodland separated by open ground or launds within which pollarded or standard trees were allowed to grow to old age.

So it was that the manors and parks slowly began to change the British landscape which still provided everything, including wood for fuel and timber for all construction purposes, for both lords and villeins alike. With increase in population, however, the latter uses became more and more important; and many of the smaller parks became no more than uneconomic status symbols and were turned over to forestry in the form of coppice management. The role of the boundary banks and ditches was now reversed, for they were refortified to keep the browsing animals out, rather than in. Protected thus, the shrubs and trees could be harvested at regular intervals of between three and twenty years. They produced *boscus*—small wood for fuel, staves, wattle, fence posts and the like—and, with rotation, even a small area of well-managed coppice could produce an annual crop and hence an annual return from the investment.

Certain trees were not cut back and harvested in this way. These were the standards which were left to grow at least towards maturity and were harvested when perhaps fifty to seventy years of age, to be used in construction work. They were replaced either by allowing regrowth of a single shoot from a coppice stool,

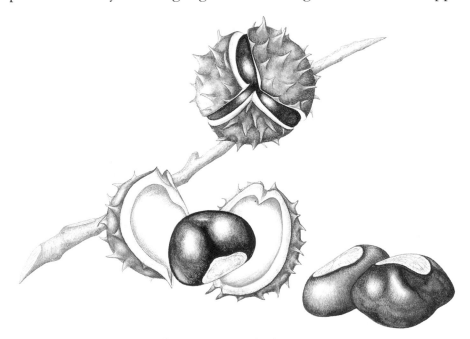

Horse Chestnuts (*Aesculus hippocastanum*)

when the resulting tree would have a distinct kink at its base, or by the planting of a maiden tree raised from seed. The standards provided *merimium*—large timber used mainly in the construction of the timber-framed houses for which the Middle Ages are so famous. The really gigantic timbers required as king posts in windmills, colleges, cathedrals and the like were mainly culled from parkland, for the standards of the ordinary coppice were not allowed to grow as big. Whether this was by design or simply due to demand, is not certain.

Domesday Book of 1086 had recorded little regarding the value of woodland, except in terms of pannage, that is the number of swine it could support. This suggests that at that time there must have been sufficient woodland for everyone to go out and help themselves to both *boscus* and *merimium*. But one has only to look at the number of medieval timber-framed houses still extant across the land to guess that such a happy state of affairs did not last all that long. As wood of all kinds became a scarcer commodity, the swine which not only ate all the acorns but rooted up the woodland scene were progressively banned, their place being taken by the beauty and productive delights of coppice.

It was during this medieval wise management that many of our 'weeds' typical of various phases of woodland development came into their own; for the coppice cycle provided a whole series of well-lit to well-shaded environments in which Primrose, Cowslip, Bluebell, Yellow Archangel, Wood Spurge, Early Purple Orchid, Bugle, Mezereon, Herb Paris and many many more could flourish. Coppice woodlands were floristic reserves par excellence, in which nothing went to waste for *loppium et chippium*, bark, branches, twigs, dead wood and even leaves and wood ash were sold.

Wise management of the woodland resource must have at least kept pace with much of the demand throughout this time, even during the sixteenth-century days of what came to be known as the Great Rebuild, when new timber-framed structures were erected so widely throughout the land. I state this with conviction for the two foremost authorities cited in Oliver Rackham's fabulous book on the subject agree that 'there was no serious shortage of timber in the sixteenth and seventeenth centuries'. Since the building of a medium-sized house would use up the annual increment of Oak growing in 290 acres of good coppice woodland, this indicates much about the British landscape during that time.

Incidentally, what evidence we do have suggests that there was not all that much more woodland in Domesday times than there was in 1946, note I do not say today. Certainly by the thirteenth century, when records started to be kept, managed woodlands were of great value which increased with time.

But 'no serious shortage' of building timber cannot be said about the already mentioned very large timbers required for special buildings. The design of the Octagon Tower of Ely Cathedral required 16 tree trunks, each 40ft long by $13\frac{1}{2}$in square. With the whole of the bishopric, and probably the whole of England—for they moved large timbers vast distances—to glean from, they could only procure 10 such trees. It would be wrong to say they had to bodge the job with the other 6, for the structure has lasted since 1340, but they certainly had to accommodate the design.

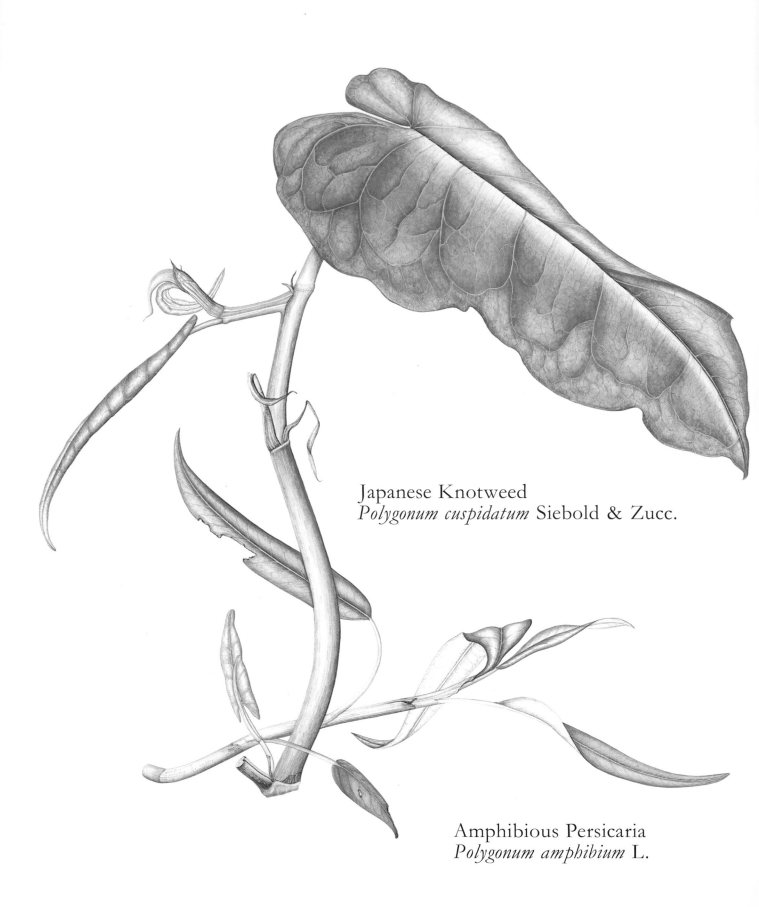

Japanese Knotweed
Polygonum cuspidatum Siebold & Zucc.

Amphibious Persicaria
Polygonum amphibium L.

(plate 24)

Ely Cathedral is the grandest piece of medieval carpentry still extant, and all would-be craftspeople should go and see it for themselves. If they can't make this ultimate sylvan pilgrimage, then they should go to the Weald and Downland Open Air Museum near Midhurst in Sussex and see other more down-to-earth structures. Wonder at the way in which they made use of all the timber unsawn, shaped by adze and used green and wayny, the bends and the bending being accommodated as the building progressed and was kept in repair. Then, as you put away your Black and Decker, think upon the fact that since 1945 half our woodland area has been converted into arable and forestry, and each minute the world is losing 150 acres of woodland and only replanting some 15 acres—much with exotic conifers and gums. What legacy, indeed what hope, are we leaving for the craftspeople of the future?

However, to return to the sixteenth and seventeenth centuries when there was no serious shortage of medium-sized building timber. Wise management was the coppice way of life in which a woodsman expected to see as much hardwood as he cut regrow in his lifetime. Such hardwood included Oak, Ash, Elm, Lime, Hornbeam, Willow and Hazel. The same probably could not be said about some of the other products of the forest, and wood for burning must have been in short supply, especially where there was local industry and in close proximity to towns. However, by that time coal was being extensively mined and imported into London by ship as was timber, the latter also from the Continent.

It was at this period, in the reign of Henry VIII which stretched across thirty-eight years (1509–47) and five-plus wives, that part of the land which is now the Queen's garden came into Royal charge. The King took over an isolation hospital situated near the village of Charing, evicting the leper maidens who had sanctuary there, and built himself a hunting box. From there he and Anne Boleyn rode out to hunt the locality, parts of which were marshy, due to the meanderings of Tyburn or over-flooding by the Thames itself. We know not which, but we do know that the whole area was eventually enclosed to form the nucleus of the Royal Park of St James, so named at least in memory of the hospital.

During the time of parks and their demise, gardens had become more common features around the houses of both small farmers and the landed gentry. They were still, in the main, simple places—fruit, vegetables and herbs being given pride of place. There were however additions, especially on the richer estates, of walkways, ponds, labyrinths, and arbours within which the owner could sit surrounded and sheltered by sylvan glory. The Wars of the Roses which started in 1455 were followed in 1461 by seemingly more settled times, and gardens expanded beyond their fortified sanctuaries. The Tudors enjoyed the formality of the art and craft of topiary, and well-clipped hedges and contrived heraldic beasts became both focal and talking points. To further accentuate such features, mounds were constructed with covered walks leading to and from them, bordered by formal flower-beds. The height of this formality were the 'knottes' in which small bushes and flowers formed geometrical designs enlivened in winter with coloured sands and stones.

The Elizabethan garden took this formality to even greater extremes with avenues of trees, raised terraces, and broad flights of steps leading down to the main garden which covered a considerable area. This was set out with a skeleton of 'forthrights'—broad gravel or paved paths leading off amongst flower-beds and covered avenues. Ponds, fountains and gigantic stone urns and vases were placed in prominent positions. The range of plants, and especially flowers, increased as new introductions from abroad began to crowd the orders of the borders. None was more orderly than Dwarf Box, which was introduced from France and used to edge the knottes.

Such formality in garden design may well have had its origins, or at least inspiration, in the orchard which had already enjoyed a long history in this land. There is good evidence that both the Greeks and Romans cultivated apples and other pomaceous fruits. It was they who first got the pip when they found that seed taken from their choicest fruit didn't necessarily breed true to form, the new tree produced in this way often being a worthless crab. So it was they who developed the basic techniques of grafting which may well have come from the east, and this knowledge was passed down through monastic scholarship. The first records of English orchards date back to early monastic times and the apples so produced were used both to eat and drink. St Brieuc when fleeing from the Saxons took eighty monks and apple stock to Normandy where they founded their cider industry, which only returned to what we like to think of as its traditional home in Devon and Somerset much later.

Throughout this time and right up to modern days the ground between the trees was kept in trim by grazing cattle or sheep. These not only kept down scrub and other plants which would compete with the apples and gave easy access to the trees both at the time of pruning and harvest, but kept the plot well-manured into the bargain. Another important part of any orchard was the admixture of different sorts of trees, and the apiary. For without the help of the bees to carry the pollen which could cross over the barriers of the recognition proteins, no fruit would be set. This formality of trees and grass may well have sparked off the idea of the garden lawn.

Whether this is true or no, it was the development of a special sort of orchard which advances our knowledge of our particular plot a little further. But before we follow the detail of this story it is as well to take a look over the other side of the Atlantic, for on 21 December 1620, the Pilgrim Fathers landed on Cape Cod, and, calling it New England, founded Plymouth which became a beachhead for colonisation.

Their detailed records tell us much both concerning weeds, the gardens they created and the ones they left at home. We read of two basic sorts of gardens, one developed by settlers who originated from the environs of Norwich, Canterbury, Colchester and London, the other developed by people from more rural areas. The basic design of both was the same, namely a compost heap which took household and garden waste fed raised beds which were contained by rough planks of wood. The former group of settlers who were much more influenced by European contact had separate beds for each main type of plant—sweet-smelling herbs in one, physic herbs for medicine in another, pot herbs in another and sallets and root vegetables in yet another—all in order. The other group who had not been subject to such continental influence back home grew their plants all mixed up together. The following is a list of the plants grown in one garden at Plymouth, all of which had been imported from England:

Barberry (Common)
Basil (Sweet)
Borage
Bouncing Bet
Catnip
Columbine
Costmary
Elecampane
Flax

Germander
Heart's-ease
Lavender
Marjoram
Mullein
Mustard (White)
Peppermint
Pot Marigold
Rue

Sage
Savory (Summer)
Savory (Winter)
Spearmint
Tansy
Thyme
Wormwood

Purging Flax (*Linum catharticum*)
and Linseed (*Linum usitatissimum*)

Wild Plum
Prunus domestica L.

(plate 25)

How many weeds were also there unrecorded and how many were brought along with the rats, mice, house flies, crows and even earthworms of later landings? The answer must be many, for fifty years later weeds causing massive trouble in the New World were legion. These included Charlock known locally as Terrify, Chickweed, Comfrey, Corncockle, Couch-grass, Dandelion, Goosefoot, Groundsel, Knotgrass, Plantain, Mallow, Mullein, Sow Thistle, Stinging Nettle, to name but a noxious few. The Barberry, which was amongst the first plants imported for hedging and for medicinal purposes, ran riot and later almost ruined the wheat crops as late as the 1920s. It can be said with truth, but with little pride, that of all the weeds which today plague North America more than 50 per cent, and the majority of the worst, originated in Europe. Having said that, it must also be said that a few came the other way; but none are of much account for though Snowberry, Gaultheria, Gallant Soldier, Michaelmas Daisy and Goldenrod can become troublesome, they can all be contained.

The reason for this mainly one-way weed traffic probably pertains to the fact that extensive agriculture has been carried out in Europe over a much longer time. This has given the process of natural selection, aided by farmers and gardeners alike, longer to work on the flora of the Old World and so produce real weeds. However, there is still time and, with modern transport speeding everything on its way, next year may see a new breakout in either continent.

Meanwhile back to the main plot of the story. Early in his reign, James I decided to go into sericulture for he came up with the novel idea that England could have a silk industry all its own. His dream was to avoid having to pay France dearly for this natural product so beloved of him and his courtiers. To this luxurious end not only did he deliver a Royal Proclamation urging all landowners to plant Mulberries, the natural food of the silkworm, but he did the same himself. With the aid of one William Stallenge he started a 4 acre silkworm farm on land between the Royal Park of St James and a scatter of fields with delightful rural names like Adam's Pasture and Upper and Lower Crow fields. The silkworms were raised in special booths and silk was produced in 'some' quantity.

That the venture lasted almost forty years must relate more to the problem of growing a crop of leafy trees than to the success of the project, for it was an economic failure. The problems were many, not the least being that Black Mulberries were planted, whose dark-green hairy leaves even the hardiest silkworm will avoid eating if it can. The choice was not entirely the fault of the entrepreneurs, but a fact of biogeography; for the White Mulberry, upon whose leaves silkworms dote, is a tree of the Mediterranean which is not well suited to the English climate. Nevertheless, both Stallenge and his son got paid for their trouble and the Mulberry Garden came into existence. In 1628 Lord Aston took over the enterprise on land close to what are now the gardens of Buckingham Palace, together with £60 per annum paid from the Royal purse to him and later to his son, to see the project on its way.

The Black Mulberries continued to flower and fruit and the silkworms continued to refuse to produce an economic amount of silk, so gradually the garden came to be used for other things. We know that Aston's son sold his disinterest in the venture to Lord Goring in 1640, and it was he who built the first large house on the site next door, where it appears on contemporary maps complete with a walled garden adjacent to a large enclosure itself labelled Goring Great Garden. Goring House was grand and its contents even grander, but as the grandiose scheme next door slowly waned, the Mulberry Garden attracted uses of a more directly pleasurable nature.

During his tenure, Lord Goring made many improvements to his lands, canalising the Tyburn first into a ditch, later into an underground aqueduct and so draining much of the marshy land. Additions to the Mulberry Garden included a long grassy tract bordered by trees, which was styled as a bowling green.

The Civil War saw both a large fort and a small redoubt built within the area though neither lasted long; it also saw the exile of Lord Goring who in 1644 became Earl of Norwich. Whatever the shortcomings of Charles I which led to the Civil War were, our British gardens have at least two things to thank him for. The first is his patronage of the Tradescants' Ark in Lambeth, without which the development of the internationally known English garden would have been greatly curtailed. Second, is his popularisation of the use of cut flowers for indoor decoration.

Ivy-leaved Speedwell (*Veronica hederifolia*)

However, history tells us that his shortcomings were more significant and in 1649 he lost much more than his interest in the Mulberry Garden, which then became just one small part of Cromwell's Commonwealth. During this sad puritanical time, the title to the grounds of Goring House and the Mulberry Garden passed through many hands and much litigation, helping at least to prove that all men and women have other than divine streaks, and all demand their rights but would rather forget their own wrongs.

Whoever was actually in charge at the time, in 1651 the Mulberry Garden was planted with several sorts of fruit trees and another area with Whitethorn (Hawthorn) in the manner of a wilderness or maze. The houses which had been designed for winging and spinning gave way to booths for wining and dining. A visit by John Evelyn of *Silva* fame in 1654 left the record that it was the 'only place of refreshment in town for people of best quality to be cheated at'. The poet William Dryden went there to eat mulberry tarts and Sir Charles Sedley wrote plays, one actually called *The Mulberry Garden*, which hinted at its less puritanical aspects. Whether the death of Cromwell and the restoration of the Crown with the fun-loving court of Charles II helped, we can only guess; but in 1663 Samuel Pepys, himself no prude, spoke of it as being 'full of people of a rascally, whoring, roguing sort, only a wilderness, though pretty'.

The Earl of Norwich returned from exile to claim his lands once more and found his claims contested at every turn. Making no headway, he sought help from his relative, one Henry Bennet later Lord Arlington, who was to broach the matter to his friend the King who was at that time, 1671, using at least part of the Goring grounds as a physic garden. It should be noted that the Worshipful Company of Apothecaries founded theirs at Chelsea two years later, where it still remains open to this day. The King must have listened with some interest, for in 1672 Henry Bennet was granted the lease and in 1677 became freeholder of the whole property, excepting the Mulberry Garden itself, which he held on a 99-year lease. By this time, however, it had lost its reputation and was closed to the public.

Fire destroyed Goring House while Arlington was in self-'exile' in Bath. He however returned to rebuild what became an even more grandiose home; so grandiose indeed that it inspired Dryden, perhaps replete with mulberry tarts, to write a long poem of adulation. In it he sang the praises of the house and its formal gardens which included everything an Elizabethan gardener would have desired and more. Their formality included terraces, steps, fountains and a greenhouse, and they overlooked fair green fields on one side and the Royal Park on the other. Unfortunately the poem, long as it is, makes no real mention of the weeds, so all we can do is guess that they were there in great profusion.

Arlington died in 1685, leaving Goring House to his only daughter, Lady Isabella Bennet who, upon the death of her husband the Duke of Grafton, rented the house and its fine gardens to the Devonshire family. But whereas Dryden had written

> While the fair vales afford a smiling view
> And the fields glitter with the morning dew
> No rattling wheel disturbs the peaceful ground
> Or wounds the ear with any jarring sound,

an account written during the tenancy of the Devonshires describes the lawns as besmirched with smuts. Poetic licence Dryden's words may be, but by that time London was expanding at an enormous rate as a city fuelled by coal. The coppice woodland around must have long since ceased to cook the meals of the metropolis, let alone heat it through winter and the coal, first dug in Newcastle by Royal Charter granted in 1239 and brought in by sea, was fouling the air, streets and people's lives with soot. In fact wood, which was recognised to be a non-polluting fuel—for though it smokes it produces no sulphur dioxide—was by then so expensive that it could only be used in the houses of the rich.

Whether it was wood or coal which started the blaze we do not know, but at this time the house was once more consumed by fire. That which remained and the land about was sold to John Sheffield, Earl Mulgrave, who soon after the accession of Queen Anne to the throne was made Duke of Buckingham or Buckinghamshire, it is not known exactly which. On the same plot, though greatly enlarged by the purchase of the Crow fields and a small slice of the Royal Park donated by the doting Queen, he built Buckingham House, facing out east across the park. Henry Wise, gardener to Queen Anne, was commissioned for the princely sum of £1,000 to landscape the new gardens, and this he did with such effect that in 1708 the whole was spoken of as a palace set within a garden which had everything— terraces, parterres, orchards, fountains, waterworks with avenues of lime— overlooking and adjacent to a meadow full of cattle. Perhaps here was the beginning of the great landscape movement which was soon to replace the excesses and excessive expense of formalism with a more natural style—indeed with the only style of garden which can be truly said to be 'English'.

Buckingham died in 1721, leaving the house and its wonderful garden to his only daughter from whom the Prince of Wales, soon to become George II, tried to buy it. However 'Princess' Buckingham, Princess because she was the natural daughter of James II by Catherine Sedley whose father had written the play *The Mulberry Garden*, asked too much for it and so lived there to the end of her life. It then passed to Charles Herbert, natural son (though not by her) of her husband who, on taking the name Sheffield, came into the house and the fortune. The former he did not find to his liking, and so sold it to George III. The plot was once again in the hands of the Monarch.

We now pass on to a period of history during which botany and the gardens of Buckingham House, which came to be known as the Queen's House, became closely linked. Princess Charlotte of Mecklenberg-Strelitz, wife of George III, was both a patron of botany and botanic gardens, and an amateur botanist of note. So much so that the *Flora Bedfordiensis*, published in 1783, was dedicated to her; and Sir Joseph Banks named the fabulous Bird of Paradise flower *Strelitzia regina* in her honour. She was no passive botanist either, for not only did she take lessons from Sir J. E. Smith, who with Banks must be thanked for saving the Linnaean collections for posterity, she was also a regular visitor to the then developing Kew Gardens. There she not only made drawings of the flowers, but amassed both through purchase and her own efforts a considerable collection of pressed plants, part of which may be seen in the herbarium at Kew today.

Indeed both the King and Queen began to spend more and more time living out at Kew. Perhaps it was to that retreat out in the country far away from bustling Belgravia which was now hemming in the Queen's House, its gardens and park that the King owed his long life and long reign, second only to that of Queen Victoria. And during this time his mother, Princess Augusta, herself a dedicated botanist, began the foundation at Kew of what were eventually to be the Royal Botanic Gardens.

The reign of George III also saw a revolution take place in the English garden. William Kent, gardener, had returned from a tour of Italy full of good ideas to do

(plate 26)

Bird Cherry
Prunus padus L.

Cherry-laurel
Prunus laurocerasus L.

Blackthorn
Prunus spinosa L.

away with the formality of the sixteenth-century garden with all its weeding and planting out. His aim was to create a more easily managed, natural style. He set to work with a will and an entirely new way at Stowe in Buckinghamshire. There he created as natural a landscape as possible, with green hills, clumps of trees and winding unobtrusive pathways reappearing at vantage points which were beset with temples. In 1740 he was joined by a young apprentice gardener by the name of Lancelot Brown.

The young man was quick to learn from Kent, Gibbs, Bridgeman and others of what came to be known as the Landscape School. He soon set up business on his own in Hammersmith, where one James Lee was busy writing a treatise on botany, and Brown soon became a master of the new art. Sweeping away all formality, removing the need for temples, grottos and terraces, he saw the capabilities of each site and used them to create a natural landscape. Natural they were, although they became classics in their time, and no less than 170 gardens soon bore the natural stamp of 'Capability' Brown.

Humphrey Repton followed on; and though not a gardener by training or by trade, he used his artistic ability and love of the countryside to become a self-styled landscape gardener—it said so on his business card—and he had equally great success. This was the time when the fortunes of our garden weeds changed dramatically. For some it was a time of retreat. Those which had thrived in flower beds, kitchen gardens, walls and gravel paths were banished to small hidden corners out of sight of the grand design. For others, like the plants of grassland and woodland edge, it was a time of new opportunity as they made themselves at home in ever-widening vistas.

The long reign of George III meant that George IV came to the throne and to the Queen's House late in life. He, however, set to with a vengeance making up for lost time; he added greatly to the house, calling it Buckingham Palace. At the same time he set about landscaping the Gardens; and to help him he employed William Townsend Aiton, son of the garden manager at Kew. Together, with much help from labourers, they remodelled the whole garden—which began to take on much of the aspect it has today—on the land which had been the Crow Fields to the west of the Palace. The work went on throughout the reign of William IV and on into that of Queen Victoria who, along with her favourite spaniel Dashy, found the Gardens to her liking. It was early in her reign—the longest of any English monarch—that the Garden Parties, which were at first called 'Breakfasts', were instituted, although they were then quite small affairs.

The most amazing fact is that throughout all this time of Royal ownership and occupancy, the Garden, in its various forms appears, until the early nineteenth century, to have lacked any really strong boundary. Fences and railings, hedges and trees there had been, for they are shown on maps and pictures and are described; but there is no firm evidence of a wall. It appears likely that the first substantial one was constructed in the late 1820s, although in 1834 boundary fences on the north and south sides were said to have been completed, with gates. In the same year the flower gardens within had to be protected with wire against the depredations of rabbits and other game; so at least up to this time there was free access to both animals and plants which came attached to the former or were just blown on the wind. In the same way waterfowl (and the list is long) could have brought seeds on their feet or feathers. The introduction and feeding of exotic birds such as the silver pheasants of which the young Queen was very proud, may well have led to a number of introductions, even as today seed fed to the Palace Garden birds continues to sow new records.

In 1838 the wall was raised by 3ft and the back gates were constructed. It was raised again in 1849 and the existing cheval-de-frise was improved after Sir John Grey saw someone climbing in from Constitution Hill. The wall was raised once again in 1872 to provide extra protection from intrusion, yet the wall which borders Buckingham Palace Road with its fine gates was not completed until 1860. Whether the wall, once completed, was any real barrier to the further entry of aliens is doubtful. However, its presence, coupled with the increasing tarmacadamisation of the surrounding area, must have cut down the incidence.

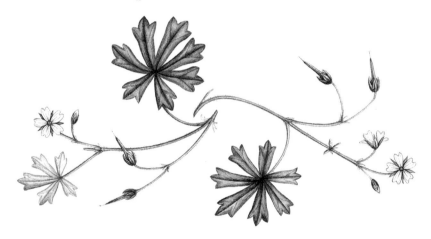

Dove's-foot Cranesbill (*Geranium molle*)

During this time other alterations had been made to afford at least some modicum of privacy if not protection for the Queen and her Garden. The Serpentine Lake became a feature of the landscape in 1828 and the material excavated from it, together with much imported from outside the Palace, was used to build the Mound. This provided winter sledging for the Queen's many children, was a boon to all ruderals while under construction, and when planted with trees at least screened the busy stables from the house. The imported material in part consisted of rubbish, but unfortunately no exact location was recorded. We know however that much gravel was imported from Hyde Park and with it must have come an influx of new seeds, perhaps even new species. The same is true when earth was imported at later dates to improve the borders and the beds, and we know that some of it came from Kensington Gardens, some from Shirley Common and some from Norwood. As both the latter are listed as containing plants of interest in De Crespigny's *New London Flora* published in 1877, it is interesting to scan those lists and surmise which of the Queen's Weeds might have arrived with soil from these localities. The following twenty-five are possibilities:

(plate 27)

Wild Cherry
Prunus avium (L.) L.

Bird's-foot Trefoil	Creeping Buttercup	Lady's Bedstraw
Bittersweet	Dandelion	Meadow Vetchling
Black Nightshade	Dog Daisy	Ragwort
Bramble	Dog's Mercury	Self-heal
Broad-leaved Willow-herb	Dog Rose	Sheep's Sorrel
Bugle	Dove's-foot Cranesbill	Tansy
Bulbous Buttercup	Fat Hen	White Mustard
Chamomile	Fool's Parsley	
Coltsfoot	Goosegrass	

Only guesswork, but important for whatever came in that way, even if already present, would mean the important possibility of cross-pollination with new genetic stock.

The history of the Lake is detailed in the records, which mainly revolve around problems of water management and of an adequate supply. That it has often been stagnant due to lack of flow, and highly eutrophic (enriched) due to the presence of a diversity of wildfowl, indicates at least one source of seeds of water plants. The addition on 20 May 1861 of lake trout and thirty-one goldfish from a hatchery in Kensington Gardens—must have further diversified both micro- and megaplankton. However such 'weed' sources fade into insignificance when it is recorded that 2,000cu yd of Thames ballast were delivered for the lake in January 1896, and seven van-loads of water plants from Kew in 1904. Man as always is the weeds' greatest friend.

The invention of the mechanical mower by Edwin Bedding in 1830 made the job of lawnsmanship much easier for, as was noted in the patent, the scythes which had been used until that time left 'circular scars, irregularities and bare patches, which continue visible for several days'—shades of hovver bovver to come.

It must be said that Buckingham Palace was not Queen Victoria's favourite residence, and when Prince Albert died in 1861 the house was left almost empty although the gardens were not left to decay. However the weeds must have had an easier time of it. Only after the Queen's Golden Jubilee in 1887 did the house and its garden take on their real role at the hub of the mightiest Empire on earth, and those Garden Parties came to the fore of society life.

There is no doubt that the most abundant plants in this Royal treasury are those which grow to make the vast area of lawn. It is made up of Bents, Smooth-stalked Meadow-grass, Rye-grass and Crested Dog's-tail. Annual Meadow-grass is abundant beneath the trees. Yarrow and Daisies do survive the mowing and trampling, but the only abundant 'broad-leaved' weed is Chamomile of the single-flowered variety, which is well looked after and allowed to flourish and spread. Similar colonies are known from the lawns of both the palaces of Kensington and Hampton Court. It has for long been both a walking and a talking point at the Garden Parties and I am sure many visitors have gone home with a bit to grace less exalted lawns. That is of course the way of weeds for, by our definition, they seize each and every opportunity and have done so ever since gardens were invented, making their way of life more profitable, at least in biological terms.

'BUCKINGHAM PALACE — PLEASE!'

If you ever want a cab in London on the day of the Royal Garden Party, my advice is keep your wife out of sight until you have your foot well in the door and the taxi meter is on. Then, and only then, reveal her plus hat, dodge the expletives, sit well back in the seats and gently say, 'Buckingham Palace, please!' If you yourself are in morning dress, it is an even more impossible task to combat the BCGPS (Black Cab, Garden Party Syndrome), the symptoms of which are taxi meters ticking over in a slow and painful manner as the immaculate occupants get more and more excited or more and more depressed, depending on the state of the weather.

Did you know that in an average year the Queen's 49 acres of London receive an incredible 26,172,459 imperial gallons of rain? Well, if you did, that long slow crawl down the Mall is not the best time to have to dwell upon it even though the cabby may well be full of bright paraphrases like, 'It won't do your hat much good, but remember its rain what makes all the flowers grow.' Of course he's right, for without those meagre few inches per average year,—for that is what all those gallons really amount to—there would be no herbs, shrubs or trees, in fact no garden at all. So, sit back and enjoy the spectacle.

As you approach the Queen Victoria Memorial, you begin to get the feeling of what it is like to be the Queen herself as you sit in state in your limousine with all the tourists peering in and waving, each one hoping that you are someone of great importance who will sit up and be recognised. At long last you are through the gates open in welcome, past all those exceedingly young policepeople and guardsmen, and are stepping down into the Palace Forecourt to join an enormous throng of equally excited and elegantly clad partygoers. Present your invitation card, pass into the cool shade of the Palace itself and then out into the full glare of a lovely sunny day, onto the Terrace, and there it is—the Queen's own back garden.

The Terrace is 137m (150yd) long and is flanked on one side by the west façade designed by Nash with additions by Blore, and on the other side by a balustrade complete with urns and vases. Though the west façade is made of real Bath stone, the balustrade and its better ornaments are made of artificial Coade stone which was synthesized to a special formula only known by Mrs Coade, and now sadly lost. This is indeed a pity, for not only is it a very beautiful medium, but it also wears extremely well even in the long-polluted air of central London. Fortunately there are some very good artificial stones now on the market which put a range of garden ornaments within the earning compass of even a modest gardener.

The Terrace is the home of Oxford Ragwort, London's commonest weed, and *Bryum argenteum*, London's commonest Moss. The former, though no more than 6in high, was flowering well on my own privileged occasion and the silvery pollution-resisting leaves of the latter edged some of the stones to perfection. The mortar was adorned in places with *Candelariella aurella* and *Lecanora dispersa*. Well, I think that's what they were, but my wife wouldn't let me bend down to get a real good look.

The façade, complete with its terra cotta frieze of Roses, Thistles and Shamrock, which at least hints of the presence of weeds amongst the Queen's flowers, must also be the home of both Starlings and Pigeons. Tell-tale feathers were there in evidence along with some with much more than a hint of rose pink. Whether the latter had arrived on the scene recycled via one of the many hats is a matter of pure conjecture, but there was no getting away from the fact that they had originally been owned by *Phoenicopterus ruber ruber*—but more of those later.

(left) Stinking Chamomile (*Anthemis cotula*), (centre) Scentless Mayweed (*Tripleurospermum maritimum* ssp. *inodorum*), (right) Scented Mayweed (*Matricaria recutita*)

The terrace is a good jumping off point to all the glories of the garden, and there could be no better landing place than the Duke's own helicopter pad complete with Daisies, Chamomile and Rib-wort Plantain. You can of course take the more olde worlde way and walk down the steps past ordered rows of capsules of London's second most common moss, *Ceratodon purpureus*.

Standing now on the main forthright, there is a choice of direction, north west or south east, enticing you to take a circular, or rather a triangular lozenge-shaped, tour around the garden. In the knowledge that the vanguard of the guests take tea before going on their horticultural perambulations, we decided to head north west to take in the Queen's borders before they got too crowded.

An herbaceous plant by definition is one that should not produce woody tissue. But, as every student of botany knows, even the common annual sunflower undergoes secondary thickening (another name for producing wood) as it grows from a seedling up towards the world record height which increases in sponsorship each year. So it stands to reason that an herbaceous border, however large, would have only small-scale plants, unless the definition of a herb is allowed a lot of expansion. What is perhaps more to the point, a border of real annual herbs would take an immense amount of annual upkeep. So, an herbaceous border has, in the accepted sense of the word, come to mean an area of prepared ground all set out with perennial plants which spring up anew each year to bear leaves and flowers, albeit on woody stems. It is also true to say that this has only recently come to be the accepted sense of the word, for the herbaceous border is an invention of the present century, though it reflects the inspiration of two great Victorian gardeners—William Robinson and Gertrude Jekyll.

William Robinson, a fiery Irishman, was both a prolific writer and doer of the gardening word. He favoured the natural look and created borders which were so filled with plants that there was little or no room for weeds and hence no need for weeding. A Robinson Border, once mature, would almost look after itself year after year.

Gertrude Jekyll, on the other hand, veered away from the natural look, producing borders which required more annual care. After a very distinguished career as an artist and embroiderer, Miss Jekyll was forced by failing eyesight to seek another outlet for her undoubted talents. She chose landscape gardening and, together with the architect young Ned, later Sir Edwin Lutyens, formed a creative partnership for which the world must be eternally grateful. Until then she had created with paintbrush and threaded needle, now she did the same with a palette of blooms which changed with the seasons. Herbaceous or mixed—it doesn't really matter. Her borders put real magic into more than three hundred gardens and were an inspiration to the whole art of gardening.

Though designed by neither of these masters, their influence is there in the Great Border of Buckingham Palace. It is, as Robinson would have desired it, on the grandest scale—170yd in length and 7yd in breadth. And, as Miss Jekyll would have wished, it changes its own kaleidoscope of colour throughout the year.

In spring, a battalion of Tulips wheels both right and left down the length of their parade ground, each blooming where and when it is told from April through

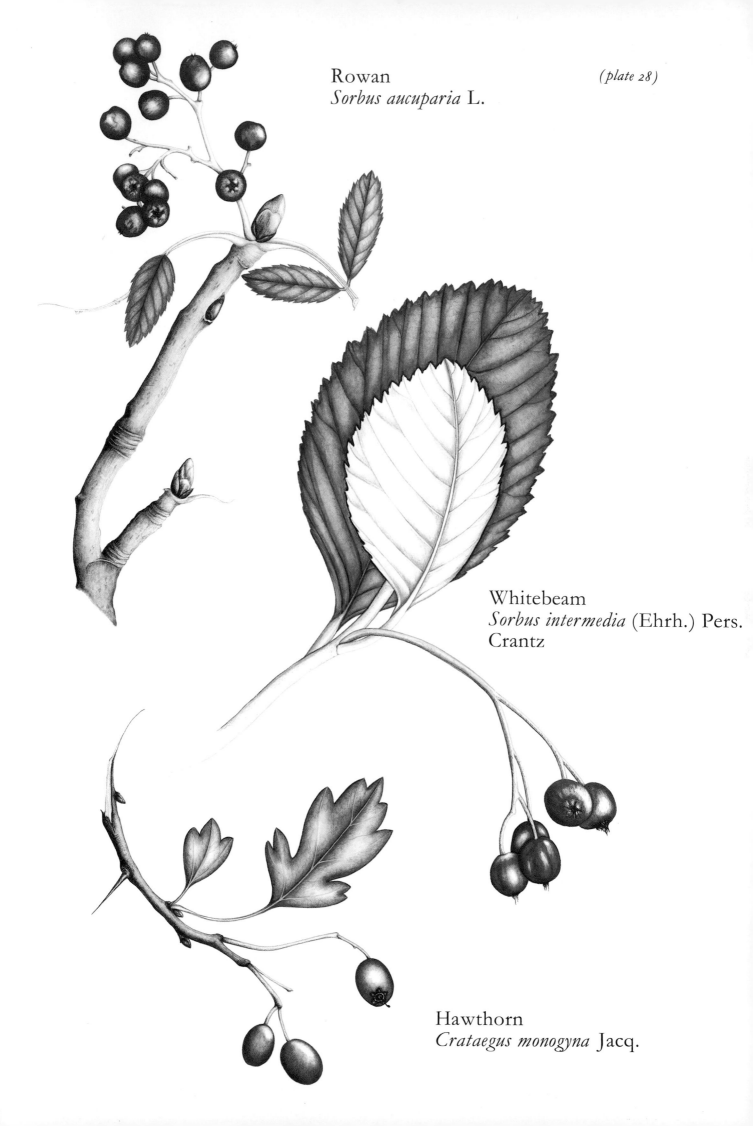

Rowan
Sorbus aucuparia L.

(*plate 28*)

Whitebeam
Sorbus intermedia (Ehrh.) Pers.
Crantz

Hawthorn
Crataegus monogyna Jacq.

May. The first Tulips bloomed in Britain in 1571, and were certainly doing well in Lambeth in 1634, for the Tradescants' catalogue for that year lists 'Tulipa, Num. 50 *Diversae* Species'.

It was in that same year that Holland went Tulip mad, so much so that the rights to the offspring of certain bulbs became a licence to print money. 'Tulip notes', as they were called, changed hands and both made and ruined many people. Though the real mania lasted for no more than three years, prices, especially for the choice varieties remained high for more than a century, too high for them to become commonplace in anything but the gardens of the rich. Today, thanks to the expertise and hard work of the horticulturalists, especially in Holland, we enjoy their stately beauty in even the humblest home.

My favourites must be the Darwins, for not only are they of great beauty and perfection, but they help to remind us that variety is not only the spice of garden life, but also of evolution. It was indeed the study of variation in nature, and the way in which both plant and animal breeders artificially select and breed what they require from that variation, which led both Charles Darwin and Alfred Russel Wallace to their momentous concept of evolution by natural selection.

In 1889, Dr. E. H. Krelage of Haarlem wrote to Francis Darwin asking permission to name a new Tulip after his illustrious father. The answer was, of course, 'yes', and the first Darwin Tulips were imported into Britain that year and have since then been selected to become the most popular of all spring garden flowers. So it is that Tulips in their guardsman's red, flame orange and golden yellow are a fitting prelude to what can only be described as a symphony of colour.

Tempting as it is to list each and every one of these swelling chords, both perennial and annual, suffice it to say there is as a background to all their tranquil beauty the martial strains of red, white and blue. As the red Tulips fade and their petals fall, the brilliant white of Shasta Daisies heralds the glorious blue of Delphiniums. There are at least thirty-six of the latter from which to choose names to keep us in patriotic mood—Blue Riband, Father Thames, Purple Prince, Duchess of Portland, Lady Eleanour and Jack Tar.

Once we could tear ourselves away from the Royal Border, every foot of which

Five-spot Burnet (*Zygaena trifolii palustrella*) caterpillar and moth on Bird's-foot Trefoil (*Lotus corniculatus*)

had something special to offer including two Holly Blue butterflies, a Five-spot Burnet and an Angle Shades Moth, we headed due west towards the great north wall, which is well screened along most of its length by magnificent trees, some old, some newly planted. Here Spotted Flycatchers, Dunnock, Greenfinch and Blackbirds were doing their best to keep out of the way of the party, although the Pigeons and Sparrows were all revved up and ready to go for the crumbs. Many good nesting sites appeared to be occupied, and there were also tell-tale wisps of straw in the shelter of the Summer House.

Everywhere trees in all stages of maturity are interwoven with meandering fairways of well-cut grass. However, between the 'mixed woodland' and the lawns are swathes which are allowed to remain in a somewhat less clipped state to accommodate the Daffodils, Jonquils, Scillas, Garden and Hybrid Bluebells and Glory of the Snow which have become at least semi-naturalised in this well-managed woodland edge. Here, too, are many of the woodland weeds like Woodrush, Speedwells and an abundance of both coarser and softer grasses.

The path now divides to enclose a central area of lawn which enshrines two large garden ornaments, both of which are to be admired. The first is a Swamp Cypress about 150 years old and doing well, though far removed from its swamp-forest origins in eastern North America. The other is the Waterloo Vase—absolutely enormous at no less than 15 ft high, it was carved from a single piece of Carrara marble and sits on thirty-six dressed slabs. Originally made for Napoleon Bonaparte, the marble having been quarried from his sister Eliza's estates, it was presented to the Prince Regent by Louis XVIII and came to London in 1815. At first it was displayed in the, then new National Gallery, until it took up its present exalted station in 1906. I dread to think what it must weigh.

The figures on the vase, be they Caesar, Napoleon or the Prince Regent—there is still much debate—look out across the stretch of grass towards four armless Tritons which support the roof of the Admiralty Temple. This came to Buckingham Palace when the land-based accommodation of Naval Headquarters had to be enlarged. Though as far removed from their natural element as the Swamp Cypress, the supporters of this bastion of the Senior Service each sit on a scaly column staring back across a sea of Roses, which themselves forge a rather special link with Napoleon's other part, Josephine. The Vicomtesse de Beauharnais, as she had been before her husband died and she married Napoleon, had at least two passions in life—the cultivation of powerful men and of Roses. As Empress, she set to with a will to transform the gardens of the Palace of Malmaison some eight miles out of Paris.

Though blessed with green fingers, Josephine did not bless France with a son and heir, and so in 1810 her marriage to Napoleon was annulled and she took up permanent residence within what soon became one of the great gardens of the world. Malmaison included amongst its many wonders no less than 250 sorts of Roses, all of which bloomed to perfection, and this perfection was captured by Pierre Joseph Redouté. His paintings now smile at us from table mats, greeting cards and calendars; as the Roses which Josephine helped to make popular now smile from gardens great and small.

(plate 29)

Cut-leaved Bramble
Rubus laciniatus Willd.

Meadowsweet
Filipendula ulmaria (L.) Maxim.

With so many from which to choose, the Queen's Garden is not the place in which to try and select a favourite. However, on such a day one may be excused a bit of name dropping, and with Napoleon's Vase not far away I will stick to those which bear French names. But where to start? Well, again on a day like this, where else but amongst the Hybrid Teas like Mme Louise Laperrière. The Hybrid Teas themselves arose out of a cross between Hybrid Perpetuals like Reine des Violettes, all noted for their size and richness of colour, and the Teas which, though more delicate than the other parent, reward us with a long and continuous flowering season. I am afraid I could find no example of the latter growing in the Garden. This could have been purely an oversight on my part, or due to the fact that today both parental types have fallen somewhat from favour, their places having been taken by their growing family of dazzling offspring.

Then there are the *rugosas*, Roserie de l'Hay double, wine red and fragrant; *damascenas*, Mme Hardy; *moschatas*, Ballerina, a pale-pink white-eyed single; *gallicas*, Presidente de Seze; *floribundas*, Chanelle, cream flushed with buff and pink and Bourbons like La Reine Victoria. These come in all shapes and sizes, from shrubs to half and full standards, their names hinting of their wild ancestors, the place from which they came, or their breeding.

Then there are the ramblers like Amethyste and the climbers which, though again bred to absolute perfection, hint much more of the unofficial wild stock which fortunately still blow unkempt about so many hedgerows. The orange double, deliciously fragrant Danse de Feu; Elegance, a *wichuriana* climber; Mermaid, the astounding bracteate climber and, back almost where we started, Mme Grégoire Staechelin, a double pink shaded crimson and oh so fragrant Hybrid Tea climber. The Rose without doubt is the Queen of Flowers, despite the prickles, although Zephyrine Drouhin has none of those at all.

Leaving the glories and order of the Rose Garden, the path leads on through shrubs and trees amongst which are another group of well-armed plants, the Hollies—*ovata* with almost unarmed leaves, *aureo marginata*, pure gold around the edges, *scotica* and *ferox*. The latter is a male clone and so does not produce fruits; it is grown purely for the fascination of its hedgehog leaves hence its common name Hedgehog Holly, and a good hedge it would make. It seems a shame that no real hedgehogs roam the palace grounds, for none have been reported. Perhaps an introduction would not come amiss.

I always like to have both hedgehogs and Holly in my garden to guard against the ravages of slugs, the former by eating them, the latter keeping them all at bay if the leaves are spread around the base of prize plants. However, as the gardens of Buckingham Palace are almost a slug desert, only four species having been recorded and all of them in small numbers, perhaps there is no need. Congratulations Mr Nutbeam!

There is one interesting fact concerning the Holly and its armour, which should be recorded here. Holly leaves are mainly armed with spines at the bottom of the tree, and often the leaves become less and less prickly above average grazing height. Whether this is due to grazers— and if so how does the plant know, or due to variation in shade from the bottom to the top of the tree, has long been a matter of conjecture. One way of settling the debate would be to keep a pet giraffe and check whether the armament of your Ilexes went more up-market.

King George IV in fact had such a pet given to him by the Pasha of Egypt. But although some of the royal Hollies may date from that time, I am afraid that the high-life browser only lived for two years, which was probably insufficient to make it a meaningful experiment. Also, in the absence of any firm knowledge concerning the length of memory of a Holly tree, I will stick my neck out no further on the matter. It is of relevance that Charles X of France was also given a giraffe at the same time and his lived for eighteen years, being fed on Rose petals, so perhaps the answer to the prickly problem is waiting there in France. Rose petals of course are modified leaves, although as far as I know they never develop prickles, however well pruned.

The path now leads to the farthest corner of the garden where one is not only close to the pulse of London traffic roaring round Hyde Park Corner, but is made to feel very close to the warm heart of the Commonwealth. There, at the foot of the great wall, are the Royal Greys. No, not the famous horses, but the plants of the Silver Grey Border, which was given to the Queen and Prince Philip to mark their Silver Wedding. A present from Lord and Lady Astor, it contains plants drawn from the furthest corners of the earth.

The silver-grey colour of all the plants is due to a felting of the translucent hairs which help to prevent the leaves and stems from being damaged by too much sun or too little rain, for they have a duel role. The silver sheen reflects the full glare of the sun and acts as a sunshade, while the weft of hairs covers the stomata and helps cut down excessive water loss. They are thus, in the main, plants from climates which are typified by annual periods of drought. In England they do well only against a south-facing wall, and as south-facing garden walls don't come much better than this Royal one, the bed is perfect for their constitution.

Lavenders, Sages, Thymes and Mints all sound as English as could be, but many of those on show in the Palace Gardens have their origin in Mediterranean lands. Each adds its own special fragrance, and thus delight to the growing army of the gardeners who are visually handicapped. Thanks to the International Year of Disabled, very many of us learned that disability is more often than not only in the mind of the able.

The scents which emanate from the leaves of the majority of these plants may well tickle both our noses and our palates, but in nature they serve a very different purpose. They are, in fact, there to ward off would-be herbivores, many of the mini-grazers finding them distasteful if not poisonous. These same plants and their kin provide essential oils for both the perfume and the chemical industries, and much research is now underway to develop them as new chemical resources for a renewable future. Plants which are able to ward off the attentions of the insect hordes may well provide chemicals which we can use as insect repellants. Lavender is probably the oldest such repellant still in use and pyrethrum, a product of a Daisy which would look well in any grey border, is a renewable safe insecticide.

Another member of the Daisy Family is present in the border and gives us a direct link with those most successful weeds of all—Groundsel and Oxford

Ragwort. It is *Senecio laxifolius*, a member of the same genus, a good common name for which would be Nelson's Ragwort. It comes from the country north-west of Nelson in New Zealand—just one of the many plants from that plant-rich country which are now finding their way into gardens throughout the world.

I saw the plant again while attending the opening of the new botanic garden in Auckland, along with so many other exciting things now being developed and bred for the world market. One of these which is already doing well in the Palace Gardens is Harakeke—New Zealand Flax or *Phormium tenax*. This was a fantastically important plant to the Maoris, for it provided tough fibres for the manufacture of everything from ropes and fishing nets to cloth. It was soon found to be at least twice as strong as its European equivalent and so became an important source of revenue to the new colony. Though it is still of economic importance, many of its uses have now been superceded by polypropylene; but one day when the petrochemicals run out it will once again come into its own, once more anchoring the world firmly to its roots in the soil. Until that time, decorative forms are finding ever wider acceptance in gardening circles. There are more than fifty varieties to choose from with red, copper-bronze, yellow and even pink variegation, there are dwarfs and giants and, though thriving in swamps, it does well even under dry exposed conditions. *Phormium tenax* is just one of many plants which New Zealand has to offer, just one small part of the Commonwealth of plants on which the future of humankind depends. And what a future it would be if only we could learn to conserve rather than destroy.

The Herald (*Scoliopteryx libatrix*) caterpillar and moth on Common Sallow (*Salix cinerea*)

Crab Apple
Malus sylvestris Miller

(plate 30)

The path continues through a wealth of trees drawn from an Empire upon which the sun never set, and beyond. Magnolias both from the New World and the Old—including *Magnolia grandiflora, M. kobus, M. sieboldii*, the ever-popular *M. soulangiana* in at least three varieties and *M. stellata*—are to be found, especially in the environs of the north end of the Serpentine Lake. Close by, the cascade brings the only sound of falling water to the garden, splashing cool green shade. It is beset with Hostas, Marsh Marigold, Azaleas, Rhododendrons, Whitebeams and an abundance of other shrubs and trees. Other waterside plants include Giant Prickly Rhubarb from South America, Weeping Willows and Royal Fern of local origin, thickets of Knotweeds from China and Japan, stands of Giant Reed from the Mediterranean, and the Common Reed which is so cosmopolitan that it has recently changed its Latin name from *Phragmites communis to Phragmites australis.*

The flowing, falling water also adds welcome oxygen to an otherwise stagnant and, in the past, problematical lake; which is made more of a problem by a large population of both visiting and resident ducks, geese and gulls which fowl the waters in more ways than one. These include the ultimate source of the pink-tinged feathers—the famous Flamingos whose origins lie, with that of the Giant Rhubarb, in South America. To complete the picture, a brace of Japanese Cranes stare fixedly from the bank. They originated in India, are cast in bronze, and stand close to a Sweet Chestnut Tree which does not spread, but is singularly twisted.

The path now divides around the Serpentine Mound which was raised in part from the material excavated for the Lake. Its flanks abound in Camellias and Rhododendrons, some of both of which originated in China and some in America, while hybrids between parents from both these now far-separated lands and ideologies are amongst the best of the cultivars.

We know now that Eurasia and North America were once joined as part of the great supercontinent Laurasia. There the ancestors of these plants came into being and, as the continents as we now know them drifted apart, the process of creative evolution clothed each land mass with a diverse mantle of living green within which the Camellias, Rhododendrons and Azaleas of both the New World and the Old went their own sweet ways. Only with the advent of the passion of gardening were they brought together again and hybridized to enrich still further our gardens and our lives. With more than 136 Azaleas and Rhododendrons and more than 80 Camellias to choose from in such a setting, and with a perfect infusion of *Camellia tea* to wash it down with, what more proof could anyone need of the potential of this Commonwealth of Plants.

THE NAME AND THE FAME

A cat sat on the mat and while sitting there it looked at the Queen, two facts which I take it still hold true, as they did in the picture book with which I was first initiated into an understanding of the written word. Written or spoken, the identification of things and persons by name is the essence of any language and of the sciences of taxonomy and classification. You cannot have one without the other.

The fact that there are a number of cats which unofficially roam the gardens of Buckingham Palace and only one Queen Elizabeth II, indicates the complexities of and the need for identification, taxonomy and classification. I am sure that no one waiting expectantly outside the Palace railings would fail to recognise the Queen, or any of the other prominent members of the House of Windsor. Likewise few of them would fail to greet the very rare sight of a cat parading across the great front courtyard with an exclamatory, 'Oooh Aar, Look, it's the Queen's Cat'. They would, however, be at least half wrong in that particular identification, for though the animal in question might well be a cat, it certainly wouldn't belong to the Queen. A Palace Garden cat or a cat-trespasser, yes; and what is more, one which probably has a name and would be recognised by its owner and his or her immediate family. However, it wouldn't be identified by, and shouldn't be identified with, Her Majesty, for the Queen prefers dogs and the palace cats if any live in, where else, the mews.

If the Queen sees one of her corgis or labradors, she will immediately be able to identify it by name, and would be able to pick it out from a pack of similar dogs. In order to accomplish this, the Queen doesn't have to go through their characteristics—long ears, greying coat, bright eyes etc—one by one and then come to a decision. She knows immediately that it is a certain dog or bitch and identifies it correctly by name.

The human eye and mind is very good at assessing a plethora of detail and coming to a correct identification. We all do it all the time. Despite this, I am often asked, 'How do you manage to recognise all those plants and remember all those long Latin names?' I always counter such a question with, 'How do you do it with all those footballers, racehorses, pop stars, titles of classical music etc etc?' For Danny Blanchflower, Hyperion III, Showaddywaddy and Rimsky-Korsakov's Third Concerto in E, Kökel something or other, are no less difficult than *Bellis perennis*. It is simply a matter of interest and familiarity not only breeding contempt, but also correct identification.

So how did it come about that the plants which surround our lives with beauty and without which we could not live, get their long Latin names, and how and why are they classified? The close association of plants with people goes back throughout antiquity, and wherever remains and artefacts of man are found, they are usually found in close association with plant material. Plants which were made use of in some way either for food, medicine, cosmetics, or in the manufacture of clothing, implements or shelter. It thus seems safe to conclude that as language developed these plants were among the first things to be given names of reference.

The first writings we have which list plants by name date from 5,000 years ago and come from Sumeria. Perhaps better known are the utterances by the King of Babylon concerning plants and urban ecology: 'From the cedar that is in Lebanon, even unto the hyssop which springeth out of the wall'. That this is better known stems from the fact that the Bible has been translated into more written languages than any other work of history.

Theophrastus, who lived in Greece during the fourth century BC is, however, usually identified as the Father of Botany though he derived many of his ideas and much of his knowledge from Plato and Aristotle. In his great work on botany he describes many plants and their parts in great detail and with great insight. For example, he recognised that the flowers of the members of what we now call the Daisy Family, consist of a head or capitulum of many small flowers. He also understood, at least in some measure, the mode of formation of the annual rings of tree trunks, and recognised order within the arrangement of floral parts. However, as far as the grouping (classification) of his named plants, he only went as far as grouping them into trees, shrubs and herbs. Nevertheless, within his writings lies the foundation of the pure science of botany.

Cato, Varro, Virgil, Galen and Dioscorides followed on in the footsteps of natural philosophy, but all were practitioners not purists. The first three were

Smooth Tare (*Vicia tetrasperma*)

Dog Rose
Rosa canina L.

(plate 31)

concerned with agriculture, and their works translated into Latin became the texts of farming across Europe. The latter two were physicians and their writings were to set the mode and the mood of botany for almost sixteen centuries. So much so, that throughout that long period the plants described by them were thought to be the sum total in existence, an idea which was to lead to all sorts of problems and to the modern legacy of 'if it's in print, it must be true'. To compound these problems the Greek Herbals as they came to be known, with their long complex descriptions and drawn illustrations, were copied by hand—a process which led to many inaccuracies, especially in the diagrams which were often drawn for artistic effect and expediency rather than botanical accuracy.

So it was that budding botanists in other parts of Europe struggled, often in vain, to fit the plants which they found growing in their locality, and which they wanted to use to cure their patients, into those early descriptions of plants, many of which did not grow outside the region of Mediterranean climate. Strange as it may seem to us twentieth-century globetrotters, the fact that different parts of the world had different floras was not generally appreciated until the sixteenth century.

The evolution of printing and the use of woodcuts for illustration both clarified and compounded the problem; for it made possible wide dissemination of information and ideas, both accurate and inaccurate. The first European text was by Otto Brunfels, who lived between 1488 and 1534 in southern Germany. Published in 1530, it contained descriptions of all the plants known to its author, listing their names in several languages, the properties which earlier writers claimed they had, and ended up with Brunfels' own judgement of 'their powers'.

As to grouping or classification, there is no attempt. The book begins with *Platago*, Plantain, 'because it is common and thus more than any other plant it bears witness to God's omnipotence'. Start with the commonest is a very practical way to begin a guide to the usefulness of plants, and a useful plant it must have proved over more than 1,600 years, for in Culpeper's most famous *Pharmacopoeia Londinensis* published 145 years later, it states not at the beginning but on page 10:

> Plantaginis Plantane. Dioscorides affirmeth that one root helpeth a Quotidian Ague, three a Tertian and four a Quartan, which though our older writers hold to be fabulous, yet there may be a greater truth in it than they are aware of, yet I am as loth to make superstition a foundation to build on, as any of them, let Experience be the Judge, and then we will weigh not modern Jurymen. A little bit of the root being eaten, instantly stayes pains in the head, even to admiration.

A very useful common plant to have growing about the garden, and a very useful book to have about the house. It was therefore no wonder that herbals began to flow from the presses almost as thick and fast as the tinctures, electuaries, decoctions, syrups and oyls flowed into willing paying patients. England's first herbal flowed from the pen and the knowledge of William Turner, who was born in Morpeth in Northumberland. It was published in three parts between 1551 and 1568.

The year 1583, four hundred years to the day before I sat down to write this treasury of common plants, was of great significance to the study of botany, for two works of great importance were published, one in Italy, the other in Holland. *De Plantis* by Andrea Cesalpino, followed the method of Aristotle almost from alpha to omega, and includes an intriguing treatise on the differences and similarities which exist between plants and animals:

> Plants feed, grow and produce offspring, but lack the sensibility and motion of animals, and therefore also need smaller organs than animals.

With decided insight, he compares the roots of one to the intestines (the feeding organs) of the other, and the shoot which produces the fruits to the reproductive system. He also does much heart-searching to find the location of such an organ within the plant, and fixes both heart and soul at the junction or collar between the root and the shoot from whence the vessels run out to supply the whole body of the plant. Quaint it may seem to us, but it must be remembered that at that time the use of lenses for magnification was in its infancy, and little or nothing was known concerning the structure of plants.

Cesalpino, too, based his classification on form—trees, shrubs, half shrubs and herbs—but added much detail concerning the fruits. He said they were equivalent to the embryo of animals, and was the first to state that the function of the leaves was to protect the fruit and that petals are modified 'foils'—active protection.

The other great botanical work of that year was The Pemptades by a Dutch writer called Rembert Dodoens. He was a physician, and though his book was subtitled *The History of Plants* it was an excellent practical treatise—an herbal pure and simple.

Without doubt the most famous—although now the facts are known, infamous—of these old herbals was that of John Gerard which appeared in 1597. It was copied almost in its entirety from a translation of the work of Dodoens, with some data from other continental authors inserted. Gerard's wife added passages of wider appeal to the ladies, and the whole was illustrated with woodcuts bought in bulk from a German printer. That much of the translation was inaccurate and that some of the illustrations were put in the wrong place was bad enough, but no mention was made as to the origin of the data or the woodcuts.

The book was however an immediate success, and made Gerard very famous, but his villainy did not stop there. He in fact used his fame to repudiate and denigrate those who pointed out his misdemeanours, even those who had helped him in his original endeavours. What is more, it is reported that he planted Peony seeds in the wild so that he could claim the discovery of a new English plant. Such was the flower power of those days that it produced this rogue for all reasons.

The sixteenth century was also embellished by the development of botanic gardens in western Europe. At first they were to be found only in the rich high-Renaissance city states of northern Italy—at Padua in 1545, Pisa in 1547 and Bologna in 1567. Later in the same century, other countries got into the horticultural act with Horti Botanici, Horti Medici, Giardinus dei Simplices,

Jardins des Plantes and Physic Gardens springing up like cress seeds at Leiden in Holland, Montpellier in France and Heidelberg in Germany to locate but a few. Britain lagged behind. Our first such institution was founded in Oxford in 1621, but what is now the most famous of all, at Kew, was not officially founded until 1842. London's first botanic garden *sensu stricto* (a place for the study of and dissemination of propagules and knowledge of plants) was opened in 1673 under the name which it still proudly bears—Chelsea Physic Garden. For a botanist or plant's-person, a visit to London is not complete until his feet have scrunched those gravel paths of garden history.

However, long before Chelsea opened its gates, London had a garden centre and natural history museum combined in Lambeth. It was called the Ark and its proprietors were John Tradescant and his son, also called John. John the Elder was born somewhere in Suffolk around 1570, and became gardener to the Cecil family at Hatfield House, to Sir Edward Wotton at Canterbury, to the Duke of Buckingham and finally to Charles I. At that time the King's garden was at Oatlands, and the Tradescants lived in a rented house in Lambeth which boasted a large garden. There they grew an ever-increasing number of plants gleaned from their travels in Europe, Russia, north Africa and the then rapidly expanding colonies of the New World. Not only were these plants on display at the Ark, but also many other curiosities both natural and man-made which they had acquired on their travels. People of substance from London and far beyond came to wonder at the curiosities, and to purchase plants for propagation in their own gardens and estates.

Spring Beauty (*Montia perfoliata*)

Just how many and which new plants the Tradescants introduced onto the British garden scene we shall probably never know, but the list certainly includes many firm favourites which are now thought to typify an English garden.

On the death of John the Younger, the collection passed as deed of gift to one Elias Ashmole, himself a collector of coins. He passed the collection on to the University of Oxford, where it formed the nucleus of the first public museum in Britain, named the Ashmolean, not as it should have been, the Tradescantian. It is said that Tradescant's wife was so disturbed by this deception, that she drowned herself in the lily pool in her garden.

In a herbal of that time, it is recorded that *Neutharis Nymphaae* Water-lilies are cold and dry and stop lust. It is unfortunate that Ashmole did not avail himself of their properties, and let the rightful name stand on that great building and head that great tradition. Tradescant is however now a household word, for it is compounded in the name of that most popular indoor plant *Tradescantia virginiana*. A more fitting name there could not be, for it was first discovered in Virginia by John Tradescant the Younger.

There is an awful lot in a name, that is if it is given and used properly. However, in 1650 the two-name system of Linnaeus which was to revolutionise both taxonomy and classification was still a long way off. Before this, classification was in the main based on utility—the uses to which the plants could be put. There was, however, a deeper reason and meaning sought by many of these earlier investigators, for they believed that through an understanding of the products of creation they could gain a deeper knowledge of the Creator himself.

Up to the period of the Renaissance, the most fundamental distinction between, and hence classification of, named things, was between those of earth and those of heaven. Of the four elements recognised by the early philosophers, earth, air and water went down, and only fire went up towards heaven. It was, however, believed that here on earth man could obtain a sense of heaven by studying the skies where perfect harmony could be found in the predictable movements of the celestial bodies. And even within the imperfections of change and decay here on earth, glimpses of heaven could be obtained from the study of beautiful creatures, works of art, saintly men and women and through public worship. This was the ideal, and the motive for able minds to enter the Church, participate in the Grand Tour or study animals and plants.

At that time, the study of animals was much further advanced than the study of plants and the zoological equivalent of the herbal was the bestiary, which focused not upon the utility of the animals to man, but on how they threw light upon human actions and destiny. It is thus not surprising that of the twelve signs of the zodiac, no less than seven are animals, so readers of magazines and watchers of Breakfast Television will know.

Aries (the Ram): Engendered with courage, self-reliance, energy, ardour, activity, generosity, strong will, bright outlook; tempered with recklessness, audacity, impatience, self-will, exaggerated self-confidence and dissatisfaction.

Taurus (the Bull): Of steadfast mind, constructive intelligence and intellect, kindliness, order, method, industry and sense of humour. But also prone to

White Water-lily
Nymphaea alba L.

(plate 32)

obstinacy, laziness, self-indulgence, delight of obstruction, parsimony and greed.

Cancer (the Crab): Full of imagination, with a retentive memory, showing constancy, loyalty, economy, tenderness to children and sympathy with old age. The following faults are however listed: morbidity, pride, tendency to brood, a miserly disposition and a pessimistic outlook.

Leo (the Lion): King among beasts, he has a large mind and warm heart, benevolence, generosity, chivalry, gratitude, free thought, tolerance, talent for organisation, power to radiate and give happiness to others. Oh, to be born between 23 July and 22 August but, wait for it, faults include self-indulgence, vanity, slackness and gullibility.

Scorpio (the Scorpion): Shows fearlessness, thoroughness, devotion in love and friendship, and is incorruptible in the face of danger and temptation. But the sting in the tail of Scorpio is vindictiveness, cruelty, jealousy, cunning, and lack of sympathy with the suffering of others.

Pisces (the Fish): Swims into power with inspiration in art, poetry and life, with romantic and lofty idealism, intuitive perception and the power of interpretation. Weight against these in the scales of his kind are negative and mediumistic listlessness, impracticality, highly strung nerves and craving for sensationalism.

Each to his own, and if you think about it the qualities and faults of each one of us can be found in the characteristics of the animal concerned, the author of each bestiary looking to the best and the worst every time to make the whole vision more universally acceptable. I myself must settle for **Capricorn** (the Goat) and I must agree that I have respect for tradition and authority, power of concentration, tact and diplomacy. As to the faults, well, this is a book on plants and we have already dwelt too long on the more mobile kind.

The Reverend Topsell stated in his *History of Fore Footed Beastes and Serpentes* published in 1678 that: 'Natural history is made by God himself, every living Beast being a word, every kind being a sentence, and all of them together a large history, containing admirable knowledge of the creator,' and missionaries and conquistadores alike found the word of God waiting for them imprinted in the nature of things.

So it was that early settlers of South America interpreted the passion-flower, its counter-clockwise tendrils representing the lash of Christ's persecutors; the three stigmas the nails; the five stamens the wounds; the corona of staminodes the crown of thorns; the ten petals the faithful apostles; and the five segments of the pedate leaf, the five fingers of our Saviour.

Signs were also sought for purposes of much more immediate use, and so it was decreed by Dr Paracelsus of Switzerland that all plants had within their form or makeup, signatures put there by God for the guidance of man. This doctrine proposed that by reading the signatures it was possible to deduce their 'powers' or 'virtues' of healing. So we find even to this day in the most up-to-date *British Flora*, Bladderwort, Heart's-ease, Lungwort, Melancholy Thistle, Navelwort, Nipplewort, Spleenwort and Toothwort. Their signatures in order are: bladder-like insect traps; heart shape of the flower; dark venous-red colour of the flower; the drooping flowerhead; the umbilicus in the centre of the leaf; shape of the flowerheads when closed in the mid afternoon; the narrow elongated shape of the frond and dark colour of the leaf stalk; toothlike shape of the rhizome and to a certain extent of the flower stalks. Of more general note: yellow flowers cure jaundice; hairy plants cure baldness; the Moonwort's crescent-shaped leaves alleviate madness; the brain-like Walnut is for afflictions of the head; while the manikin-like root of the Mandrake is a sure cure for just about all the ills in the book. It was tinctures of such facts and fancies distilled throughout the ages which sent plant-hunters on expeditions across the then known world to seek fame, fortune and envy; their fortitude and foresight opening up the richness of new lands for colonisation.

Europe as the centre of civilisation offered much to those who were—through breeding, skill, good fortune, luck or even guile—removed from the ranks of the peasantry into the status of craftsman, artisan, the professions or even the landed gentry. The peasants gleaned their daily bread and looked to heaven for the fringe benefits of a good harvest and a prudent landlord. The latter being more assured of a livelihood could look towards God in other ways—through learning, study or the Church—the ultimate in the seeking out of these gifts of grace being the Grand Tour of Europe.

The Grand Tour was indeed the unforgettable experience, wherein the privileged minority might see the whole cross-section of the attributes of civilisation, the arts and architecture, the works of saintly people and the places of learning and of corporate worship. Despite the rigours of such a trip, and there were many, this travelling apprenticeship, for that is what it was, allowed even the least able participants to gain a certain aura of expertise concerning these things which allowed glimpses of heaven. It was also good fun, for between those chinks in the celestial armour were all the desires of the heart and the devices to make them all come true.

Tobacco (*Nicotiana* sp.)

Europe was however already full to overflowing, plagued with poverty, mass illness, bigotry and persecution, so much so that even the best harvests fell short of everyone's expectations. So it was that the New World drew like a magnet, for it offered freedom from oppression, land for the taking, and livelihoods for the making; with new plants being discovered on those then far shores. Tobacco, Potato and Corn, to name but three, promised that no one need live by bread alone;

Lime
Tilia europea L.

(plate 33)

but what of the grand experience, the seeking for God in another way? A knowledge of the arts and the art treasures of the old way of life meant little in the new colonies, but a knowledge of creation, of animals and plants, was not only of spiritual but of practical benefit.

The gaining of such knowledge posed its problems, for though it was easy to take herbals and bestiaries on the voyage, even the most able descriptions of many European plants had little meaning on the other side of the Atlantic. Imagine the problems of sitting down with a splitting toothache and a book which starts with Plantain, and looking for something, anything, which would alleviate the pain. Black Hellebore, Bear's-foot or Christmas Flower, hooray, helps toothache when held in the mouth: has no flowers but shiny pedate leaves—and the nearest thing you find is Poison Ivy. Oh well, you can't win them all, and the allergy caused if you were indeed susceptible would certainly make you forget the toothache.

In those days it was therefore impossible to become a good field botanist by use of simple books of reference. It was essential to have a grand apprenticeship, visiting as many of the masters and botanic gardens as possible, and making reference to their living and dried collections. This was impossible in the New World.

The settlers of course learned much from the local Indians and it is of great interest that the first Thanksgiving party thrown by the Pilgrim Fathers in 1621, less than a year from the date of their landing, included Cranberries on the menu. Cranberries get their name, or so it is said, from the shape of their flowers which resemble the head of a crane in miniature; and it must have been easy for anyone well versed in British plants to see the resemblance between the robust vines of the sweet-sour American Cranberry and the smaller, creeping European plant which grew even within the bounds of London until the mid-nineteenth century. But, what of all the others, where could any budding New World botanist hope to make a start? What was needed was a system of classification which provided a key that took you straight, or at least straighter, to the plant of your investigation—a ray of light in the jungle of New World green. And, what is more, it had to be written in American. Thank goodness it wasn't just Culpeper who was annoying academia with translations of their rites and rituals from the Latin!

John Wilkins indeed popularised the discoveries of Galileo throughout England in his *Discoverie of a World in the Moone*, published in 1638 and *A Discourse Concerning a New Planet* in 1640, the new planet being Earth, which blasphemously went round the Sun, thereby downgrading our position at mission control to one of mere denizens of a satellite spaceship. However, he realised the problems of such vulgar discourse, and came up with the brilliant idea of a written 'language' which could be understood and read by all literate people, their brains perceiving the visual data and processing it into the vernacular they spoke. Oh joy, a language which would make science universally understood, but only through the written word and hence progressing mainly through recorded argument. It was at this very time that scientific societies first came into being, and the printing of first journals and later proceedings and transactions containing scientific papers which had been discussed at their meetings became both the boon and the bane of science.

To locate Wilkins within the time-frame of both science and history, he married Cromwell's sister and was contemporary at Oxford with Robert Boyle, Robert Hooke and Christopher Wren, architect of the Ashmolean Museum. To locate plants and animals and other things within the sphere of reference of his new 'experesso', for it would be wrong to call it a language, Wilkins used symbols and a classification worked in units of three. The system came to nought but the Ray of hope which emerged to enlighten world botany was called John.

John Ray was born, probably in 1627 at Black Notley in Essex, eldest son of a well-to-do blacksmith who was prosperous enough to send his heir to Trinity College, Cambridge. Wilkins had been appointed Master of that college in 1660 and he invited Ray to do the botanical part of his treatise. Ray soon found that not all of nature worked in threes, but with classification in mind went on to greater things after a period in the Church and a Grand Tour which provided him with plant specimens enough to keep him hard at work for the rest of his life. In his *Historia Plantarum Generalis,* a vast work of 2,860 folio pages which summarised the entire botanic knowledge of that time, he used the following basis of classification though modified through later works and reprints.

The division between trees, shrubs and herbs was there as was the recognition of the important distinction between those plants which had but one seed leaf, and those which produced two. All these fundamental distinctions which were handed down from Theophrastus are still in use today. Following Cesalpino, Ray made much of the structure of fruits and seeds but linked with it the importance of the form of roots, shoots, leaves and flowers in order that, as he said, plants which resembled each other could be grouped together. Although Aristotle had used the ideas of family and genus (the latter meaning relation), Ray was the first to seek and understand the relationships and to talk about natural families of plants. Ray was also abreast with the findings of both Malpighi and Nehemia Grew who were at that time using microscopes to investigate the anatomy of plants, and he used many of their findings in his system.

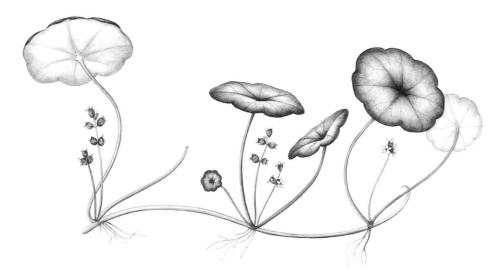

Marsh Pennywort (*Hydrocotyle vulgaris*)

Sub-groupings were numerous, fining down to the genera, each of which was given a name and then a short description. The species into which the genera were divided were then characterised in a few words followed by a more detailed description. Groups of like plants divided into genera and species was a practical system for practical use; but it was much more than that for the author had great insight into the workings both of nature and of nature's God. Ray believed that plants belong to the same species if they arise through seed from a plant similar to themselves. He also knew that the number of species was invariable for on the seventh day God rested from all his labours.

However he also states that different coloured flowers should not be regarded as separate species for the colour variation could not be maintained through the seed, but only through cuttings. He thus describes and makes much of variation within the species and, if that is not enough, believed that invariability of species was not absolute for variation can take place through degeneration of the seed. To uphold the latter belief he quotes a number of examples including the fact that from seeds of *Primula veris major* (now known as Primrose Cowslip hybrids) *Primula pratensis inodora* have arisen—a fact which even today gives rise to much wonder wherever meadow flowers are still allowed to bloom.

If only he had taken more note of Grew's growing evidence of sex in plants, how much further might this brilliant blacksmith's son have taken the science of botany. Ray went to his death-bed in 1705, safe in the knowledge he had painstakingly written in *The Wisdom of God manifested in the Works of Creation* published six years before.

Joseph Pitton de Tournefort who was contemporary with Ray but worked in France, held much closer to Cesalpino's dictum that fruits and seeds were most important in classification. He in fact adds flowers to the list of important characters, and attempts to find evidence as to why the other plant parts are of less importance. His fine and accurate descriptions of many genera are still in use today. He went still further in his classification, grouping the plants he described into a number of classes, which are again aggregated into several sections. However, he makes less of the species, almost brushing them aside with notes on stem and leaf.

During his Grand Tour he had seen date-growers carry flowering branches from 'male' palms and suspend them amongst the fronds of the 'female', thereby securing a rich crop of fruits. This act had, of course, been recorded by Herodotus in Babylonia, and Theophrastus had himself formed a notion that all other plants were likewise male and female:

> In Trees, considered universally and taking in each several kind, there are, as has been said, many differences. One of these is common to them all, namely that by which they are distinguished into Female and Male, of which the one bears Fruit, the other not, in some kinds; in those which both bear Fruit, that of the Female is the best, unless these are to be called Males, for so they are by some.

Just why it took the philosophers and scientists some 2,064 years to expand upon these points we will never know. It is, however, now a fact of history that from 370 BC to AD 1694, it was a case of no sex please we're botanists, despite the fact that between those dates the palmists had read between the lines and cashed in with a harvest of rich sticky fruits every year.

It was left to a German botanist working in that famous university of Tübingen finally to prove beyond all doubt that plants had sex. His name was Rudolph Jacob Camerarius, and though a medic by profession, he cultivated a large number of different types of plants and found that if the male flowers were picked off in time, there would be no fruit, while fruit would certainly develop if the pistils of the female flowers were provided with pollen.

So it was that sex raised its beautiful (for in the guise of a flower it could never be regarded as ugly) head within the science of botany—a fact which was to revolutionise both classification and the science itself, making it of much wider appeal and application. Thus we find a letter from a physician by the appropriate name of Alexander Garden who had for two whole years struggled to make some sense of the flora around Charleston, South Carolina, where he had taken up practice in 1752:

> I have wasted two years in following Tournefort's system, in which I was first instructed by Dr. Charles Alston in the Edinburgh garden. Having been invited to Carolina, (where for some time I have, with tolerable success, practised medicine) being furnished with the *Institutiones Rei Herbariae* and the writings of Ray, I made daily excursions into the country. But the immense labour of reducing the plants I collected into proper orders, to say nothing of the uncertainty attending the investigation of genera and species, and, still more, of determining varieties, according to the Tournefortian system, was all so very tiresome, that at length my patience was exhausted, and had I not by good fortune, met with your most excellent works mentioned above, I might have been stopped in my progress and have altogether given up this most pleasing of pursuits.

The letter was to one Carl Linné, better now known to the world under the latinisation of Carolinus Linnaeus, and the above-mentioned work was his *Systema Naturae*, first published in 1735.

His father, Nils Ingemarsson, though a peasant who lived in the province of Småland in Sweden, was destined for the priesthood and, as was the custom of those days, any children born to him would have been christened something or other Nilsson. However, while at the local village school he adopted the family name of Linnaeus after a mighty Linden tree which grew near his home. The Linden was regarded as a sacred tree by the country folk and in 1704 Linnaeus was ordained as priest at Råshult, where he grew and studied many herbs in his large garden. His first son Carl was born on 23 May 1707 and even from his earliest childhood was a keen botanist and gardener.

At school, Carl did not excel in the sciences or the humanities for, as he later

(plate 34)

Columbine
Aquilegia vulgaris L.

said: 'Crude schoolmasters in a crude manner gave the children a mind for the sciences enough to make their hair stand on end.' But despite such shock-horror methods his physics teacher recognised within young Linnaeus a great gift for natural science, provided him with the works of Tournefort and urged his family to let Carl study for medicine rather than for the Church.

At the universities of Lund and Uppsala all was not plain sailing, mainly due to lack of money. His nascent genius however showed through and, thanks to sponsors and a teaching post in the botanic gardens, he continued his studies and went on collecting-expeditions to both Lapland and Dalecartia. He travelled to Holland in order to obtain the paper qualification Doctor of Medicine, which would allow him into practice, and graduated at the University of Harderwijk in a week—a week which included enrolment, preliminary examinations, the printing of his thesis which was already written, and passing his viva voce examination. Knowing his faith and his beliefs, I am sure that on the seventh day he too rested.

By this time he had little money left and so in search of work travelled to Amsterdam, then Leiden, where under the patronage of a number of wealthy people he published his *Systema Naturae* at the tender age of twenty-five. Publications of even greater note followed on apace and his growing fame took him to France and England before he returned to his beloved Sweden, where eventually in 1741 he became the Professor of Botany at Uppsala. There his genius and sheer hard work brought students from all over Europe to follow him in ever-increasing numbers around the botanic gardens which he reorganised to become one of the finest in Europe. And he put both Uppsala and botany firmly on the scientific map, for his advice concerning questions of natural science was sought by private individuals and governments alike. So it was that Alexander Garden wrote from America, and such was Linnaeus's interest and capacity for work that he immediately entered into correspondence with him to help him in his New World endeavours.

It was in the formal naming of plants that Linnaeus did so much, for up to the publication of his *Species Plantarum* in 1753 the Bulbous Buttercup which still grows in abundance in the Queen's Garden would have been described in classical Latin as *Ranunculus foliis ovatis serratis, scapo nudo uniflora* (Ranunculus with oval-toothed leaves, with a naked stalk bearing a single flower). From that date forward it was termed *Ranunculus bulbosus* L. and as such was known exactly for what it was across the world of science. The first name signifies the genus which embraces some 300 yellow or white flowers now known to grow across the world; the second name—which in modern usage should always begin with a small (lower case) letter—denotes the species, the accepted characters of which should as stated by Ray always breed true through its seed.

The problems of this and many other definitions of species, mainly concern the characters of the plant thought to be important in species delimitation. Throughout the history of plant taxonomy there have been both 'splitters' who regard any difference however small as reason for specific identity and so another name, and 'lumpers' who do exactly the opposite. It is perhaps more fun and certainly more prestigious to be a splitter for the newly created name, *Ranunculus bulbosus* L. for example, bears the authority in a capital letter (L. stands for Linnaeus) and so immortalises the name both of the plant and of the botanist who first recognised the uniqueness of its characters and split it off from the rest. The type specimen, identified and named by him, is pressed and mounted on a sheet of fine white paper before being given pride of place and kept in safety in an herbarium.

It was also the custom to name plants after patrons, famous people, colleagues etc. Alexander Garden was thus immortalised by Linnaeus in the generic name *Gardenia*; Professor Rudbeck, who had first befriended him at Lund, in *Rudbeckia*; while *Boerhaavia* is named after Professor Boerhaave, who was a patron in the publication of his first book. It therefore might be thought that the name of Linnaeus would be borne by a plant of great magnificence; but no, it is commemorated in one of the least significant plants to be found in the Boreal forests which in Linnaeus's time still swathed much of his homeland—the modest twin-flower *Linnaea borealis* L. His own description of the plant includes the following: 'A little northern plant, flowering early, depressed, abject and long overlooked'. Some writers have said that this is an apt summary of the man and his career, but on this I cannot agree. A little northern plant, flowering early, yes; but with his first major work published at the age of twenty-five and a prestigious chair of botany at thirty-four, the rest is far from the truth. There is, however, little doubt that his name raised this little flower to greater eminence than it perhaps deserved, for it became part of the crest of the ennobled family von Linné. One of the pupils he sent to China sent back a tea-service of finest porcelain, its only decoration being this flower and, when the Cardinal of Noailles erected a cenotaph in his garden to the memory of the naturalist, he planted the *Linnaea* by its side as the most appropriate ornament. King Gustavus III of Sweden pronounced the solemn eulogy.

White Campion (*Silene alba*)

(plate 35)

Traveller's Joy
Clematis vitalba L.

Meadow Buttercup
Ranunculus acris L.

Winter Aconite
Eranthis hyemalis (L.) Salisb.

White Dead-nettle
Lamium album L.

Hedge Woundwort
Stachys sylvatica L.

Red Dead-nettle
Lamium purpureum L.

(plate 36)

Such was the man; but apart from his twin-flower and his two-name system of identifying plants, what did he do for posterity and for those would-be botanists struggling to understand nature's God out in the colonies? Through the use of the sexual system he put a key within the grasp of anyone who could make simple observations, using a hand lens. And to demonstrate this, I quote with modification from *An Introduction to Botany* by James Lee, nurseryman at The Vineyard, Hammersmith. Published in 1788, this work contained the first full exposition in English of the Linnaean system.

> The Sexual System was invented by Dr. Linnaeus, Professor of Physic and Botany at Upsal. It is founded on the Parts of the Fructification. (By Fructification we are to understand both the Flower and Fruit of Plants, which cannot well be separated. For though the fruit does not swell and ripen until after the Flower is fallen, its Rudiment of first Beginning, is in the Flower of which it properly makes a part.)
>
> It has been said, that the Pollen was destined for the Impregnation of the Germen. This is performed in the following manner. The Antherae which at the first opening of the flower are whole, burst open soon after, and discharge the Pollen, which dispersing itself about the flower, Part of it lodges on the Surface of the Stigma, where it is detained by the Moisture with which that part is covered, and each single Grain of the Pollen bursting and dissolving in this Liquor, as it has been observed to do by the Microscope, is supposed to discharge something that impregnates the Germen below. What the substance is that is so discharged, and whether it actually passes through the Style into the Germen, seems yet undetermined, it being difficult to observe such minute Parts.

We now know what it is and indeed any child with a simple microscope, a drop of sugar-water and patience, can observe the rapid growth of pollen tubes. However, in the eighteenth century there was still much to be learned, but this new knowledge was sufficient to point Linnaeus to take count of the right features which put both Orders and Classes within his system. Lee goes on to state that:

> By the Sexual System, Plants are disposed according to the Number, Proportion and Situation of the Stamina [male parts, filaments and anthers] and Pistillia [female parts, germen, styles and stigmas]. A new set of Principles have been derived from them, by means of which the Distribution of Plants has been brought to greater Precision, and rendered more conformable to true Philosophy in this System, than in any one of those which preceded it. The Author of it does not pretend to call it a natural one, he gives it as artificial only, and modestly owns his Inability to detect the Order pursued by Nature in her Vegetable Productions. But of this he seems confident, that no natural System can ever be framed, without taking in the Materials, out of which he has raised his own, and urges the necessity of admitting artificial Systems of Convenience, till one truly natural shall appear.

The latter still holds to this day, for as yet no fully natural system has ever been agreed upon and many of even the most popular modern books published on plants do little to really help the beginner to get to grips with putting a name to the plant he has found in the field.

So with no more ado, we will outline the Linnaean system as it applies to the weeds of Buckingham Palace, and indeed the whole Treasury of Common Plants which follows will be cross-referenced in this way. All you need do to take the first steps towards identification is to recognise and count the parts—the public parts—of any flower; for arrayed within such beauty to attract the attention of passing insects they cannot be called private.

Ground Ivy (*Glechoma hederacea*)

LINNAEAN SYSTEM OF CLASSIFICATION

The genera of the Queen's Weeds are classified with some modification according to the sexual system of Linneaus as developed and expounded in the English language by James Lee in *An Introduction to Botany* (1788). An outline of the complete classification is given, although members of all Orders are not represented within the Queen's Lists.

To use the classification you must know what a perfect flower is, and as all flowers are perfect in the eye of the beholder, especially if he or she grew them, we must define perfection in the botanical sense. A perfect flower is one that has both male parts (anthers, consisting of filaments and pollen sacs) and female parts (pistillia, consisting of ovaries with styles and stigmas). Figure 1 which is reproduced from *An Introduction to Botany* may help you to understand these their Privy Seals; the numbers indicate the twenty-four Linnaean classes. Anthers are usually easy to count, but when it comes to the pistillia you may experience a little more difficulty.

The number of pistils is usually taken to be the number of stigmatic lobes, for these are usually the most obvious. However, in those flowers which do not have separate pistillia but have them fused, the number can be determined by slicing that part of the flower across, and counting the number of chambers within. It is these chambers which will eventually hold the seeds.

Foxglove
Digitalis purpurea L.

Common Figwort
Scrophularia nodosa L.

(plate 37)

PLANTS WITH ALL FLOWERS PERFECT

CLASS 1   Plants whose flowers each have but 1 stamen

ORDER 1   Flowers each with 1 pistil
ORDER 2   Flowers each with 2 pistils

CLASS 2   Plants whose flowers each have 2 stamens

ORDER 1   Flowers each with 1 pistil
Genera: Ash (*Fraxinus*); Privet (*Ligustrum*); Speedwells (*Veronica*); Gipsywort (*Lycopus*); Sage (*Salvia*).

ORDER 2   Flowers each with 2 pistils
Genus: Sweet Vernal-grass (*Anthoxanthum*).

ORDER 3   Flowers each with 3 pistils

CLASS 3   Plants whose flowers each have 3 stamens

ORDER 1   Flowers each with 1 pistil
Genera: Iris (*Iris*); Crocus (*Crocus*); Spike-rush (*Eleocharis*); Club-rush (*Scirpus*); Bulrush (*Schoenoplectus*); Galingale (*Cyperus*).

ORDER 2   Flowers each with 2 pistils
Genera: Reeds (*Phragmites*); Pampas Grass (*Cortaderia*); Purple Moor-grass (*Molinia*); Sweet-grass (*Glyceria*); Fescue-grasses (*Festuca*); Darnel (*Lolium*); Meadow-grasses (*Poa*); Cock's-foot Grass (*Dactylis*); Dog's-tail Grass (*Cynosurus*); Brome-grasses (*Bromus*); Couch-grass (*Agropyron*); Yellow Oat (*Trisetum*); Soft-grasses (*Holcus*); Hair-grass (*Deschampsia*); Bent-grasses (*Agrostis*); Foxtails (*Alopecurus*); Canary-grass (*Phalaris*).

ORDER 3   Flowers each with 3 pistils
Genus: Spring Beauty (*Montia*).

CLASS 4   Plants whose flowers each have 4 stamens, all of which are of equal length (if unequal see Class 14)

ORDER 1   Flowers each with 1 pistil
Genera: Buddleja (*Buddleja*); Plantains (*Plantago*); Dogwood (*Thelycrania*); Vervain (*Verbena*); Bedstraws (*Galium*).

ORDER 2   Flowers each with 2 pistils
Genera: Field Madder (*Sherardia*); Fluellen (*Kickxia*); Monkey Flower (*Mimulus*); Parsley Piert (*Aphanes*).

ORDER 3   Flowers each with 4 pistils
Genus: Pearlwort (*Sagina*).

CLASS 5   Plants whose flowers each have 5 stamens

ORDER 1   Flowers each with 1 pistil
Genera: Ivy (*Hedera*); Primrose (*Primula*); Yellow Loosestrife (*Lysimachia*); Pimpernel (*Anagallis*); Periwinkle (*Vinca*); Duke of Argyll's Tea-plant (*Lycium*); Nightshades (*Solanum*); Tomato (*Lycopersicon*); Tobacco (*Nicotiana*); Mullein (*Verbascum*); Bindweed (*Convolvulus*); Bellbines (*Calystegia*).

ORDER 2   Flowers each with 2 pistils
Genera: Dodder (*Cuscuta*); Chervil (*Chaerophyllum*); Beaked Parsleys (*Anthriscus*); Knotted Parsley (*Torilis*); Earthnut (*Conopodium*); Water Dropwort (*Oenanthe*); Fool's Parsley (*Aethusa*); Hogweed (*Heracleum*); Marsh Pennywort (*Hydrocotyle*); Elms (*Ulmus*); Goosefoots (*Chenopodium*).

ORDER 3   Flowers each with 3 pistils
Genus: Elder (*Sambucus*).

ORDER 4   Flowers each with 4 pistils

ORDER 5   Flowers each with 5 pistils
Genus: Flaxes (*Linum*).

ORDER 6   Flowers each with more than 5 pistils
Genus: Horse Chestnut (*Aesculus*).

CLASS 6   Plants whose flowers each have 6 equal stamens; if the 6 are not of equal length, see Class 15

ORDER 1   Flowers each with 1 pistil
Genera: Lily of the Valley (*Convallaria*); Fritillary (*Fritillaria*); Bluebell (*Endymion*); Grape Hyacinth (*Muscaria*); Snowflakes (*Leucojum*); Snowdrop (*Galanthus*); Daffodil (*Narcissus*); Rushes (*Juncus*); Wood-rushes (*Luzula*).

ORDER 2   Flowers each with 2 pistils

ORDER 3   Flowers each with 3 pistils
Genus: Sorrels and Docks (*Rumex*).

ORDER 4   Flowers each with 4 pistils

ORDER 5   Flowers each with more than 4 pistils
Genus: Water Plantain (*Alisma*).

CLASS 7   Plants whose flowers each have 7 equal stamens

ORDER 1   Flowers each with 1 pistil

ORDER 2   Flowers each with 2 pistils

ORDER 3   Flowers each with 4 pistils

ORDER 4   Flowers each with 7 pistils

CLASS 8   Plants whose flowers each have 8 stamens

ORDER 1   Flowers each with 1 pistil
Genera: Sycamore (*Acer*); Heather (*Calluna*); Heaths (*Erica*); Willow-herbs (*Epilobium*); Rosebay Willow-herb (*Chamaenerion*).

ORDER 2   Flowers each with 2 pistils

ORDER 3   Flowers each with 3 pistils
Genus: Persicarias (*Polygonum*).

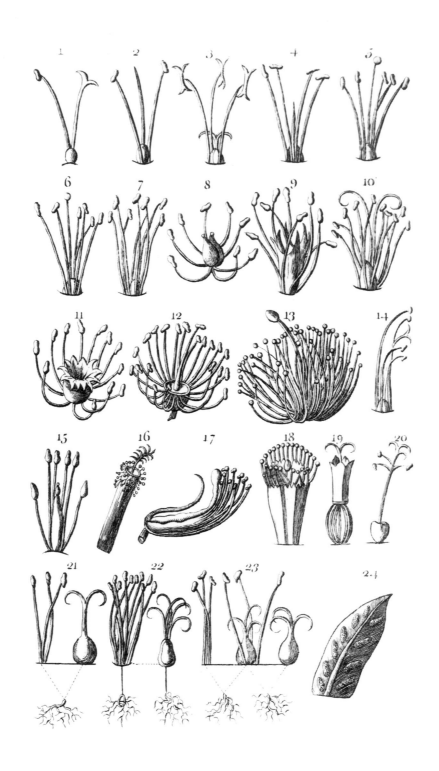

Figure 1. The Privy Seals from *An Introduction to Botany* (1788)

ORDER 4   Flowers each with 4 pistils

CLASS 9   Plants whose flowers each have 9 stamens

ORDER 1   Flowers each with 1 pistil

ORDER 2   Flowers each with 3 pistils

ORDER 3   Flowers each with 6 pistils

CLASS 10   Plants whose flowers each have 10 stamens

ORDER 1   Flowers each with 1 pistil
Genera: Strawberry Tree (*Arbutus*); Rhododendrons (*Rhododendron*).

ORDER 2   Flowers each with 2 pistils
Genus: Saxifrage (*Saxifraga*).

ORDER 3   Flowers each with 3 pistils
Genera: Campion (*Silene*); Stitchworts (*Stellaria*).

ORDER 4   Flowers each with 5 pistils
Genera: Common Mouse-ear Chickweed (*Cerastium*); Spurrey (*Spergula*); Sorrels (*Oxalis*).

ORDER 5   Flowers each with 10 pistils

CLASS 11   Plants whose flowers each have 12 to 20 stamens

ORDER 1   Flowers each with 1 pistil
Genus: Purple Loosestrife (*Lythrum*).

ORDER 2   Flowers each with 2 pistils

ORDER 3   Flowers each with 3 pistils

ORDER 4   Flowers each with 5 pistils

ORDER 5   Flowers each with 12 pistils

CLASS 12   Plants whose flowers each have 20 stamens or more arising from the rim of the calyx, the fruit developing below them

ORDER 1   Flowers each with 1 pistil
Genus: Plums, Blackthorn and Cherries (*Prunus*).

ORDER 2   Flowers each with 2 pistils
Genus: Hawthorn (*Crataegus*).

ORDER 3   Flowers each with 3 pistils
Genus: Whitebeam and Rowan (*Sorbus*).

ORDER 4   Flowers each with 5 pistils
Genera: Meadowsweets (*Filipendula*); Apple (*Malus*).

ORDER 5   Flowers each with more than 5 pistils
Genera: Cinquefoils (*Potentilla*); Rose (*Rosa*); Brambles (*Rubus*).

CLASS 13   Plants whose flowers each have more than 20 pistils arising from the rim of the calyx, the fruit developing above or within them. A footnote warns

unlucky 13: 'The fruits of this Class are often poisonous, which makes it necessary to distinguish them from those of the last, which abounds with eatable fruits'

ORDER 1   Flowers each with 1 pistil
Genera: White Water-lily (*Nymphaea*); Limes (*Tilia*).

ORDER 2   Flowers each with 2 pistils

ORDER 3   Flowers each with 3 pistils

ORDER 4   Flowers each with 4 pistils

ORDER 5   Flowers each with 5 pistils
Genus: Columbine (*Aquilegia*).

ORDER 6   Flowers each with 6 pistils

ORDER 7   Flowers each with more than 6 pistils
Genera: Winter Aconite (*Eranthis*); Traveller's Joy (*Clematis*); Buttercups (*Ranunculus*).

CLASS 14   Plants whose flowers each have 4 unequal stamens, the two outermost being longest

ORDER 1   Flowers each producing 4 seeds, each in a separate coat, and the pistillum is forked
Genera: Mint (*Mentha*); Self-heal (*Prunella*); Woundwort (*Stachys*); Horehound (*Ballota*); Dead-nettles (*Lamium*); Hemp-nettle (*Galeopsis*); Ground Ivy (*Glechoma*); Bugle (*Ajuga*).

ORDER 2   Flowers each producing more than 4 seeds which are enclosed together in a dry fruit
Genera: Figwort (*Scrophularia*); Foxglove (*Digitalis*).

CLASS 15   Plants whose flowers have 6 unequal stamens each, the 2 opposite ones being shortest

ORDER 1   Flowers produce a fruit, a short roundish pod or pouch
Genera: Wart-cresses (*Coronopus*); Shepherd's Purse (*Capsella*).

ORDER 2   Flowers produce a fruit which is a long, many-seeded pod
Genera: Cabbage and Chinese Mustard (*Brassica*); Mustard and Charlock (*Sinapis*); Radish (*Raphanus*); Bittercresses (*Cardamine*); Yellow-cresses (*Rorippa*); Hedge Mustard (*Sisymbrium*); Thale Cress (*Arabidopsis*).

CLASS 16   Plants with flowers which have their stamens combined in one set by fusion of at least part of their filaments

ORDER 1   Flowers each with 3 stamens

ORDER 2   Flowers each with 5 stamens

ORDER 3   Flowers each with 8 stamens

ORDER 4   Flowers each with 9 stamens

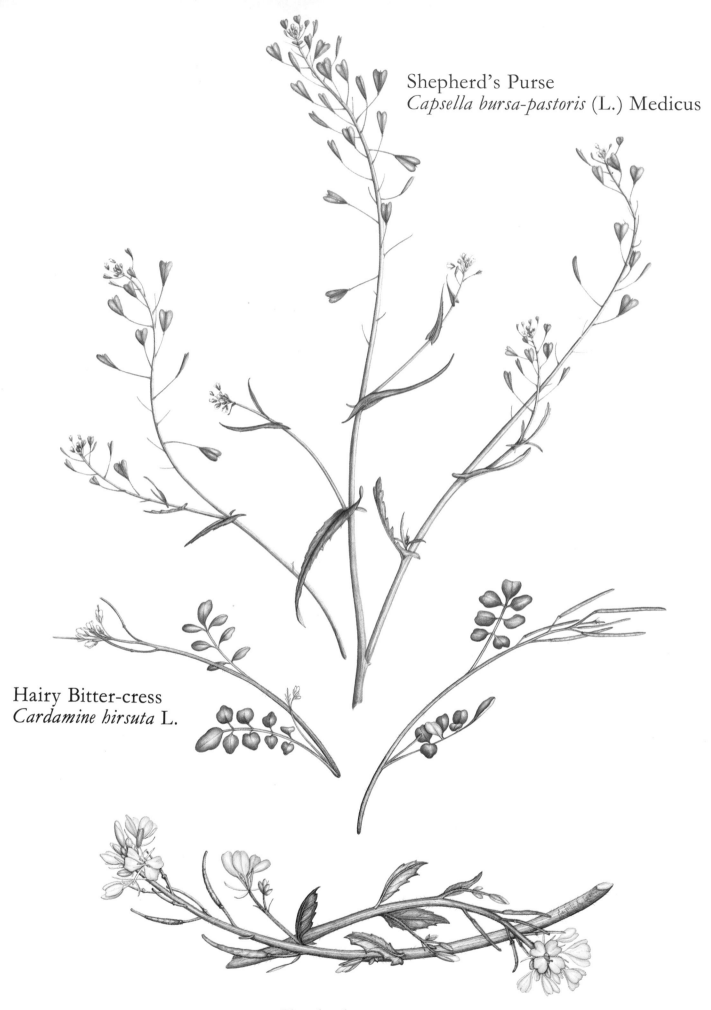

Shepherd's Purse
Capsella bursa-pastoris (L.) Medicus

Hairy Bitter-cress
Cardamine hirsuta L.

Charlock
Sinapis arvensis L.

(plate 38)

ORDER 5   Flowers each with 10 stamens
Genus: Cranesbill (*Geranium*).

ORDER 6   Flowers each with 11 stamens

ORDER 7   Flowers each with 12 stamens

ORDER 8   Flowers each with more than 12 stamens

CLASS 17   Plants whose flowers have their stamens combined in 2 distinct sets (brotherhoods)

ORDER 1   Flowers each with 5 stamens

ORDER 2   Flowers each with 6 stamens
Genus: Fumitories (*Fumaria*).

ORDER 3   Flowers each with 8 stamens

ORDER 4   Flowers each with 10 stamens
Care must be exercised here, for all these plants do not have their stamens in two distinct sets. However, the shape of the flower which is like that of a Sweet Pea is enough to overcome any mistakes, for all those in Class 16 have regular flowers without the irregularities of keel and wings. Those with only one set (brotherhood) of stamens: Laburnums (*Laburnum*); Brooms (*Sarothamnus*)

Those true to the main characters of the class, each flower having 2 sets (brotherhoods) of stamens: Clovers (*Trifolium*); Bird's-foot Trefoil (*Lotus*); False Acacia (*Robinia*); Bird's Foot (*Ornithopus*); Vetch (*Vicia*); Peas (*Lathyrus*).

CLASS 18   Plants whose flowers have their stamens combined into more than 2 sets (many brotherhoods)

ORDER 1   Flowers with 5 stamens in each set

ORDER 2   Flowers with 12 stamens in each set

ORDER 3   Flowers with 20 stamens in each set. None represented in these Royal Lists, but this order includes the Oranges and Lemons

ORDER 4   Flowers with more than 20 stamens in each set
Genus: St John's Worts (*Hypericum*).

PLANTS WITH SOME OR ALL FLOWERS IMPERFECT

CLASS 19   Plants whose flowers have their stamens united into a tube, and are borne in compound heads called capitulae

ORDER 1   All the florets in the compound head are perfect, that is they have both male and female parts, 5 stamens, 1 pistil and producing one seed

Genera: May be divided into two groups:
  1   Those whose individual florets are all ligulate (strap-like) and hence not unlike a single petal: Nipplewort (*Lapsana*); Cat's Ear (*Hypochaeris*); Hawkbit (*Leontodon*); Goat's-beard (*Tragopogon*); Lettuce (*Lactuca*); Sow-thistle (*Sonchus*); Hawkweeds (*Hieracium*); Hawk's-beards (*Crepis*); Dandelion (*Taraxacum*).
  2   Those which have definite tubular flowers, but be careful because the florets have a petal-like limb: Burdock (*Arctium*).

ORDER 2   Plants with compound flowers in which those in the centre of the head are perfect, while those around the radius bear only female parts
Genera: Again they may be considered in two groups:
  1   Those in which all the florets are tubular: Cudweed (*Gnaphalium*); Wormwood or Mugwort (*Artemisia*).
  2   Those which have ligulate ray florets around the radius of the head: Gallant Soldiers (*Galinsoga*); Niger (*Guizotia*); Ragworts (*Senecio*); Coltsfoot (*Tussilago*); Butterbur (*Petasites*); Fleabane (*Pulicaria*); Golden-rod (*Solidago*); Michaelmas Daisy (*Aster*); Canadian Fleabane (*Conyza*); Daisy (*Bellis*); Chamomile (*Chamaemelum*); Stinking Chamomile (*Anthemis*); Yarrow (*Achillea*); Scentless Mayweed (*Tripleurospermum*); Ox-eye Daisy (*Leucanthemum*); Scented Mayweed (*Matricaria*); Thistles (*Cirsium*).

ORDER 3   Plants with compound flowers in which the florets of the central disc are perfect, while those of the radius are neuter having neither male nor female parts
Genus: Knapweeds (*Centaurea*).

ORDER 4   Plants with compound flowers in which the florets of the central disc are male and those of the radius female

ORDER 5   Plants with compound flowers made of many partial flowers all contained within a single calyx

ORDER 6   Plants which have simple flowers, with their stamens united into a tube
Genus: Violets (*Viola*).

CLASS 20   Plants the flowers of which have the stamina either growing upon the style of the pistillium or upon a receptacle which stretches out in the form of a style and supports both the stamina and the pistillium

ORDER 1   Flowers each with 2 stamina

ORDER 2   Flowers each with 3 stamens. Prickly Rhubarb (*Gunnera*).

ORDER 3   Flowers each with 4 stamens

ORDER 4   Flowers each with 5 stamens

ORDER 5   Flowers each with 6 stamens

ORDER 6   Flowers each with 8 stamens

ORDER 7   Flowers each with 10 stamens

ORDER 8   Flowers each with 12 stamens

ORDER 9   Flowers each with more than 12 stamens
Genera: Sweet Flag (*Acorus*); Lords and Ladies (*Arum*).

CLASS 21   Plants which have no perfect flowers, but bear both separate male and female flowers on the same plant

ORDER 1   Each male flower has 1 stamen
Genus: Spurge (*Euphorbia*).

ORDER 2   Each male flower has 2 stamens
Genera: Sedges (*Carex*)—several.
ORDER 3   Each male flower has 3 stamens
Genera: Sedges (*Carex*)—majority; Plane (*Platanus*)—may have up to 5 stamens; Reedmace (*Typha*).

ORDER 4   Each male flower has 4 stamens
Genera: Mind-your-own-business (*Helxine*); Nettles (*Urtica*); Birch (*Betula*); Alder (*Alnus*).

ORDER 5   Each male flower has 5 stamens

ORDER 6   Each male flower has 6 stamens

ORDER 7   Each male flower has 7 stamens

ORDER 8   Each male flower has more than 7 stamens
Genera: Hornbeam (*Carpinus*); Hazel (*Corylus*); Beech (*Fagus*); Sweet Chestnut (*Castanea*); Oak (*Quercus*); Water-milfoil (*Myriophyllum*).

ORDER 9   Male flowers each with one set of united stamens
Genus: Scots Pine (*Pinus*). Note the placing of a conifer here amongst the flower-bearing plants.

ORDER 10   Male flowers each of which have free stamens, but with the pollen sacs (anthers) united
Genus: Melon (*Cucumis*).

ORDER 11   Male flowers each with a false pistillium from which the stamens grow

CLASS 22   Plants which have no perfect flowers, but bear both male and female flowers on separate plants

ORDER 1   Each male flower has 1 stamen

ORDER 2   Each male flower has 2 stamens
Genus: Willows (*Salix*)—some.

ORDER 3   Each male flower has 3 stamens
Genus: Willows (*Salix*)—some.

ORDER 4   Each male flower has 4 stamens
Genus: Willows (*Salix*)—some.

ORDER 5   Each male flower has 5 stamens
Genus: Willow (*Salix*)—some.

ORDER 6   Each male flower has 6 stamens

ORDER 7   Each male flower has 8 stamens
Genus: Poplar (*Populus*).

ORDER 8   Each male flower has 9 stamens
Genus: Mercurys (*Mercurialis*).

ORDER 9   Each male flower has 10 stamens

ORDER 10   Each male flower has 12 stamens
Genus: Tree of Heaven (*Ailanthus*).

ORDER 11   Each male flower has more than 12 stamens fixed to the calyx

ORDER 12   Each male flower has many stamens not joined to the calyx

ORDER 13   Male flowers have one set of united stamens
Genera: Yew (*Taxus*)—another misplaced conifer.

ORDER 14   Male flowers have stamina whose filaments are free, but their anthers are joined

ORDER 15   Male flowers have the stamina growing out of an imperfect pistillium

CLASS 23   Plants which have both perfect flowers complete with male and female parts, and also either male or female flowers or both

ORDER 1   Plants which exhibit such polygamy on the same plant
Genus: Orache and Purslane (*Atriplex*).

ORDER 2   Plants which exhibit such polygamy on separate plants
Genus: Holly (*Ilex*).

ORDER 3   Plants which exhibit such polygamy on 3 or more separate plants

PLANTS WHICH HAVE NO FLOWERS

CLASS 24   Plants without flowers whose method of sexual reproduction was then still waiting to be discovered

ORDER 1   The Ferns and their allies, which bear their 'fruit' on the backs of their leaves
Genera: Horsetail (*Equisetum*); Royal Fern (*Osmunda*); Bracken (*Pteridium*); Hart's-tongue (*Phyllitis*); Wall Rue (*Asplenium*); Male Fern and Broad Buckler (*Dryopteris*).

ORDER 2   Mosses

ORDER 3   Algae

ORDER 4   Fungi

Easy isn't it? All you have to be able to do is count from one to twenty, and you can put your plant in its class and thus take the first step towards its identification.

If this pre-occupation with the number of stamens smacks of chauvinism, do remember that class distinction isn't everything, indeed here it is only the first step. It is consideration of the female attributes which puts order into the whole thing and gets you down towards the genera, all a step in the right direction towards understanding and classification.

No wonder Alexander Garden and many others wrote with such appreciation, despite the fact that explicit in the thesis of Linnaeus was proof positive that plants went in for all sorts of polygamy with twelve to nineteen husbands in the same marriage. The goings on amongst the flowers began to make the habits of the birds and bees seem tame in the extreme. Angiosperm, the name now reserved for flower-bearing plants, may to the purist mean 'marriage hidden in a vessel', but the facts of life were now out for all to see, to understand, sex pistils and all. However, the eighteenth century was not a period of great prudery and the sexual system was accepted on its merits. Some contemporary ladies, like Jane the daughter of Cadwallader Colden, made use of the sexual system; she herself classified and illustrated a number of plants found growing around her home in New York State and Linnaeus named a genus of plants *Coldenia* in her honour. It had 5 stamens and 1 pistil and was thus put in Class 5, Order 1, amongst a group of plants all of which have 'a naked-mouthed and pervious corolla'. But other, less scientific, classifications were on offer as alternatives—for example an Alphabet of Lovers' Flowers.

This had evidently first developed in the harems of the Ottoman Empire, where the ladies had little else to do and where lovers could not display their preferences in a more open way. Instead, flowers either single or in bunches, each signifying something special, were used in place of words. This floral code was first brought to the attention of the English in a letter written by Lady Mary Wortley Montagu from Constantinople in 1718. One can only wonder if either she or her correspondent ever read the works of Linnaeus. The floral code really came into its own with the publication of the *Language of Flowers* in 1820, and was to flourish in the days of Victorian moral restrictions.

SOME LOVERS' FLOWERS FOUND IN THE QUEEN'S GARDEN

Apple	Temptation	Lilac	Forsaken
Birch	Gracefulness	Mulberry	Wisdom
Chestnut	Render me justice	Nettle	Slander
Dock	Have patience	Oak	Hospitality
Everlasting Pea	Lasting pleasure	Poppy	Oblivion
Forget-me-not	True love	Rhododendron	Danger
Gooseberry	Anticipation	Small Bindweed	Obstinacy
Hawthorn	Hope	Thistle	Austerity
Iris	Message	Vernal Grass	Poor but happy
Jonquil	Desire	Willow Herb	Pretentiousness
Kingcup	I wish I was right	Yew	Sorrow

It is quite easy to see how these particular virtues came to be associated with at least some of these plants. All you have to do is use your imagination. Or is it just that these associations are now so much part of our English language heritage that they always spring to mind.

But apart from such fancies in which Linnaeus had no part, there were some, like John Ruskin, who later objected to the robust botany of angiosperms and polygamy. He even insisted that his students draw flowers without their private parts, though he was ridiculed both for his lack of scientific objectiveness and for his prudery. But his scruples had advocates in the staid thinking of later Victorian England, though the expurgated system of classification he advanced did nothing to denigrate the sexual system of Linnaeus.

There were, however, two points which blighted the otherwise revered name of Linnaeus and both stemmed from the same problem—heredity. In 1763, Linnaeus sought permission from his government to appoint his successor at Uppsala. This was gladly given, a fact which may seem strange to us in these 'enlightened' days, but was quite in order at that time. He appointed his only son. Just how unwise this decision was and how useless Carl Von Linné the Younger turned out to be is demonstrated by the fact that, on Linnaeus's death, his claim to his father's collection, library and correspondence was contested by his mother and sisters. What is even more distressing is that in the ensuing period of legal argument, these priceless documents of science were locked away unattended, where both rats and mould did their best to corrupt. It is perhaps unkind to say that fortunately Carl the Younger died at an early age, but his effect on the fame of the name of his family was no doubt responsible, at least in part, for the sale of those collections lock stock and type specimens, to Sir Joseph Banks, then President of the Royal Society of England.

There is little doubt that the King of Sweden was sorry to see them leave the country, but he did not, as is publicly believed and commemorated by paintings, send a gunboat in pursuit of these priceless treasures of science. There is, however, no getting away from the fact that they were of inestimable value—part of the scientific heritage of Sweden and

Mossy Pearlwort (*Sagina procumbens*)

Tufted Vetch
Vicia cracca L.

Broom
Sarothamnus scoparius (L.)
Wimmer ex Koch

(plate 39)

the world. Sweden's loss thus became Britain's gain, for in the hands of the very rich young naturalist, Sir James E. Smith, who bought them from Banks for 1,000 guineas, they became the reference point for the founding in 1788 of The Linnaean Society of London. There the vast bulk of them remains to this day in Burlington House, London, at the headquarters of the Linnaean Society.

The other matter of heredity, which damaged the name of Linnaeus to a much greater extent, was certainly not due to any action of his own, except perhaps his simple faith in God. He believed that there was order in the natural world, and that that order had been created by God; and so it was that by the middle of the nineteenth century, the name of Linnaeus was caught up in the anti-evolution wrangle. The anti-Darwinists, who claimed solidarity with the angels no less and who abhorred the theory of descent, quoted Linnaeus as their leading authority on the doctrine of Creation. It does little credit to the new biologists that they responded by attacking Linnaeus himself, and denigrating the man and his work almost out of hand.

However, once the main furore and bigotry of pro- and anti-Darwinism had laid itself to rest, the name of Carl Linnaeus flowered once more as the *Princeps botanicorum* of his youth; for 1907, the bi-centenary of his birth, was celebrated throughout the then civilised world as a red-letter day in the annals of human culture.

So perhaps the description of that 'little northern plant, flowering early, depressed, abject and long overlooked' is apt, for it does encapsulate our memory of the man himself. Its night-scented and delicate pale-pink petals openly invite both bees and botanists alike, to look upon four stamens and one pistil—the true beauty of ordered sex. This bold, shy beauty must however fade before the efficacy of hidden monogamous intercourse becomes apparent in the swelling of the fruit, a one-seeded nutlet.

Angiosperm, the fact now contained in the name, was hidden from Linnaeus. We now know that, due to the mode of formation of that part of the Germen which he could not see, each new generation in sexually reproducing organisms, be they Linnaea or Linnaeus, differs from the last and all the attributes of the father are not passed down even to his only son. There is much more in a name than meets the eye, and there is much more in a species than those characteristics which do appear to breed true through seed. There is above, below and subtending all variation, variation enhanced by the recombination of genetic information during fertilization. It is upon this innate variation that the process of natural selection acts, bringing about the evolution of novelty—and novelty is the spice of life.

The arguments still rage, and modern-day taxonomists and geneticists armed with giant computers and data which would really make Linnaeus's hair stand on end, still try to seek the true order of a natural classification.

Linnaeus sleeps, oblivious of what came after, safe in the knowledge of created order, the study of which shaped his life. We, too, will look no further than his naming of names, and will use his key to unlock the secrets of a Treasury of Common Plants.

TREASURY OF
COMMON PLANTS

Being a simple list arranged in Linnaean order of all the native and naturalised plants which have been recorded in recent times within and upon the walls of the gardens of Buckingham Palace.

Within each Linnaean class and order, the arrangement of the genera follows that set down in the *Flora of the London Area* 1983. However, in a few cases other arrangements have been followed to accentuate family relationships. It is of great interest to see that many Linnaean groupings are similar to those determined by modern taxonomy.

The information appended to each entry in these Royal Lists celebrates the fact that 'a weed is but a plant whose virtues have not yet been discovered'.

CLASS 2: ORDER 1
OLIVE FAMILY OLEACEAE
Ash *(plate 1)*
Fraxinus excelsior L.

A strong pliable wood noted for its straight grain and its plastic (bending) properties when steamed; hence used for the handles of all sorts of tools, especially scythes, and the shafts of carts. In Norse mythology ash is the tree of the Universe.

Its roots run in three directions: one to the Asa gods in heaven, one to the Frost Giants and the third to the underworld. Under each root is a fountain of wonderful virtues. In the tree, which drops honey, sit an eagle, a squirrel and four stags. At the root lies the serpent Nithhöggr gnawing it, while the squirrel Ratatöskr runs up and down to sow strife between the eagle at the top and the serpent.

In woodlands its delicate compound leaves which unfurl very late cast but a light shadow, and hence harbour a rich ground flora.

The tree was long thought to have mystical properties and was much used in medicine. Even in the late eighteenth century, young ash trunks were split and the two halves were held back by two strong men. A naked child would then be passed through the cleft, in the dark before dawn. The trunk was then bound and plastered with mud. If the tree survived and grew strong it was believed that so would the child.

The aerodynamic ash keys may be pickled according to a recipe in the *Queen's Closet*, a famous seventeenth-century cookery book, and are very pleasant to eat.

Privet *(plate 1)*
Ligustrum vulgare L.

A plant which is attractive to butterflies and moths, especially if allowed to flower. A common plant of wet woodlands, the one with the neat oval leaves which is much planted as garden hedges is an introduction from Japan.

While on the subject of this family, we must remember the Olive itself which is both of Biblical and gastronomic acclaim. Also a weed, the Jojoba is found growing in the Sonoran Desert of North America. Like the Olive, its fruits produce an oil, which is the perfect substitute for the highly valued sperm-whale oil. Crops of the plant are already a feature of California, Mexico, Israel and Australia; and there is now no longer any excuse for hunting sperm whales.

FIGWORT FAMILY (see also pages 138, 141 and 161)
Speedwells
Veronica sp.

The name Veronica derives from the saint of that name, derived in turn from the Greek *Hiera Eikon* = a sacred picture. The flowers are said to bear a representation of the face of Jesus, as was the saint's handkerchief which was long preserved in Rome; hence the French, *Véroniques* or Holy Faces. With such a signature of authority it is not surprising that it has a reputation of speeding a hundred ailments well.

Thyme-leaved Speedwell
Veronica serpyllifolia L.

Wall Speedwell
Veronica arvensis L.

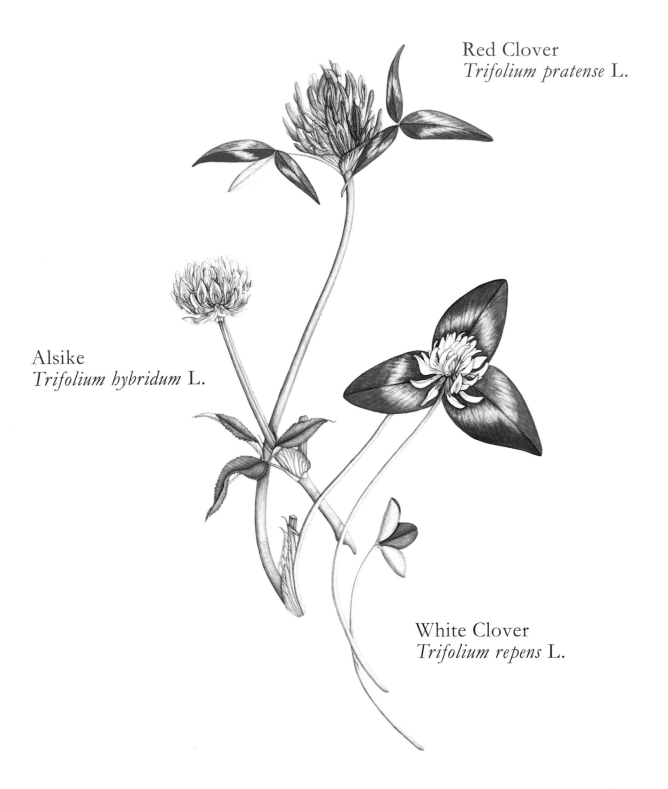

Red Clover
Trifolium pratense L.

Alsike
Trifolium hybridum L.

White Clover
Trifolium repens L.

(plate 40)

Ivy-leaved Speedwell
Veronica hederifolia L.

Buxbaum's Speedwell
Veronica persica Poiret

The first three are natives, the latter was introduced in 1825. It is of interest that Britain's two commonest Speedwells, the Common and Germander, are not found in the Royal Garden.

Speedwells may be used to make a pleasant tea which was highly prized in Germany and Sweden. Perhaps Samuel Pepys, who sent for a cup of the then new and so expensive China tea in 1661, refreshed himself on an infusion of these herbs when he visited the Mulberry Gardens two years later.

MINT FAMILY (see page 158)   LABIATAE

Gipsywort *(plate 2)*
Lycopus europaeus L.

A waterside plant with lobed and toothed leaves lacks the pungent smell of the true mints. It produces a brown-black dye which was evidently used by people who wanted to be recognised as gipsies, hence its common name.

Sage
Salvia × superba (S.-Taroula & Schn.) Stapf.

No deception here, a plant with a truly wonderful flower and taste, although that found in the Palace Gardens is an introduced hybrid.

CLASS 2: ORDER 2
Sweet Vernal-grass *(plate 3)*
Anthoxanthum odoratum L.
(A member of the Grass Family, see this page)

Along with Cock's-foot, this is probably the commonest grass in Britain and Ireland. It certainly is the sweetest to chew, for that is how it got its name. The sweetness is from a substance called coumarin which gives the smell of new-mown hay.

CLASS 3: ORDER 1
IRIS FAMILY   IRIDACEAE
Yellow Flag *(plate 4)*
Iris pseudacorus L.

The roots of this handsome waterside plant were used as a cosmetic and to cure aching teeth, and a distillation from the whole plant was prescribed for eye complaints. However, care must have been taken, for the juice is very acrid and can burn and blister. Unless you are a real expert your local GP is always the best resort.

Crocus
Crocus spp.

Over the years many cultivated types of Crocus have been planted out en masse for naturalisation in the Palace Gardens. The enormous variety which we can now enjoy is thanks to the painstaking work of the plant breeders and to the genetic variation of the wild populations originally taken in and constantly replenished from the field. The only native crocus to grow in Britain is the Meadow Saffron, which is a member of the Lily Family.

Common Spike-rush
Eleocharis palustris L.

A good plant at the water's edge, giving cover for small birds and also food in due season.

Sea Club-rush
Scirpus maritimus L.

Though identified with the sea, it is a plant of tidal estuaries; but as the Thames is tidal up to Tide-end-Town, this Royal Garden is not far from its natural habitat and its seeds were probably brought in by water birds.

Bulrush
Schoenoplectus lacustris (L.) Palla

This is the real bulrush which is used for rush-bottoming chairs and by coopers for caulking casks. It was also used as a strewing rush to cover floors and as wicks for tapers to light dark rooms.

Galingale
Cyperus longus L.

Now a rare plant, found naturally in only a few places in Britain, one of only two members of this mainly sub-tropical and tropical genus of this family to occur here. Its root furnishes a highly nutritive and agreeable food once the bitterness has been removed by drying, washing or fermentation. In the tropics both the roots, rhizomes and the seeds of other species provide food for the human population.

CLASS 3: ORDER 2
GRASS FAMILY

The Daisy Family is the most diverse and successful family in the world, and the grasses with their six hundred genera and probably ten thousand plus species, come a close second. The Queen's Weeds reflect this situation to perfection with forty-one species of the first, and twenty-eight species of the second, in their family lists.

However, in terms of area covered, the grasses win roots down in the Royal Garden, and as far as their importance on a world scale, they win fruits up. Bread and pasta are made from flour which comes mainly from wheat – a kind of grass. Corn on or off the cob, and cornstarch or oil which are in most of the foods we buy (look on the back of the packet) are, or come from, the seeds of another member of this family. The same goes for rice. These three plants which were once no more than weeds growing in the first forest clearings made by primitive peoples, now produce the staple food for a large part of the human population.

What would breakfast be like without the packs of grass seeds (cereals) cunningly manipulated into a variety of shapes and forms? Bread, biscuits, cakes all come the same way. As for the evening drink, whether you are a beer connoisseur or a whisky or gin tippler, the alcohol and much of the body (which is about to become yours) originated in the food stored by grass seeds.

If you are a complete Megan, that is a total carnivore, you are still totally dependant upon plants for that is the origin of meat, milk, butter and cheese. There are large tracts of the world which cannot, and other large tracts including much of Britain which should not, be ploughed up and put to cereals or other crops. They should be left to grass, which will feed animals, which will in turn feed us.

Anyone who calls himself or herself a humanitarian should support all those people who are actively lobbying to stop the inhumanities of factory farming. It is stupid and wrong to bring cereals from Third World countries to fatten livestock for the First World. It is also just as stupid to plough good rangeland and to grow barley or corn for feedlot, meat or milk production. We must all have faith in the grasses and their unique properties of growing in habitats in which trees and other more robust plants can't make anyone a living.

The following grasses grow in the gardens of Buckingham Palace and must be given pride of place in these Royal Lists. Sir Walter Raleigh may well have introduced Tobacco, Corn and many other amazing botanical wonders of America to Britain. He may also have laid out his cloak so that Elizabeth I did not get her feet muddy. It is, however, these common grasses which have provided this plot of land with a living carpet fit for any Queen.

Reed
Phragmites australis (Cav.) Trin. ex Steudel

A grass which covers enormous areas of tidal and freshwater marshland and edges of lakes and rivers. Of immense importance in stabilising the edges of our waterways and providing a habitat for waterfowl, other birds and insects. These marshes also provide a rich harvest of reeds for thatching and could with a little imagination form the basis of a fine papermaking industry.

Pampas Grass
Cortaderia selloana (Schultes & Schultes f.) Aschirson & Graebner

A native of South America, pampas grass is best managed by burning once it is established in your garden. This gets rid of all those dead brown leaves and nourishes the young shoots and fabulous plumes of flowers. It also reminds us that fire is part of nature's own management of the natural grasslands from which the pampas originated and upon which much wildlife and many people depend for their livelihood.

Purple Moor-grass
Molinia caerulea (L.) Moench.

A taste of real moorland in the centre of London. This is one of the plants called Early Bite by the hill shepherds, for its young shoots feed the lambs in early spring.

Great Water-grass *(plate 3)*
Glyceria maxima (Hartman) Holmberg

If you want delicate beauty floating on and growing from the water, this will help. It also provides cover for the fish and a launch-pad for Damsel- and Dragonflies.

Fescues
Festuca spp.

Rye-grasses
Lolium spp.

Meadow-grasses
Poa spp.

It is mixtures of these three types of grass which produce our lawns, greenswards and bowling greens.

Meadow Fescue
Festuca pratensis Hudson

Tall Fescue
Festuca arundinacea Schreber

Red Fescue
Festuca rubra L.

Sheep's Fescue
Festuca ovina L.

Perennial Rye-grass
Lolium perenne L.

Hybrid Rye-grass
Lolium hybridum Hausskn.

Annual Meadow-grass
Poa annua L.

This is the commonest grass of London's pavements and is always in flower.

Wood Meadow-grass
Poa pratensis L.

Rough Meadow-grass
Poa trivialis L.

Cock's-foot
Dactylis glomerata L.

Readily recognised by the 'cock's-foot' shape of its tall flowerhead.

Crested Dog's-tail
Cynosurus cristatus L.

The flowerhead of this looks like a dog's tail, one side of which has been flattened by wagging on the carpet.

Soft Brome-grass
Bromus mollis L.

Wonderfully hairy.

Couch-grass
Agropyron repens (L.) Beauv.

I can find little good to say about this particular plant, except that it composts well and so returns the minerals it stole from deep in the soil last year.

Yellow Oat
Trisetum flavescens (L.) Beauv.

Yorkshire Fog
Holcus lanatus L.

Creeping Soft-grass
Holcus mollis L.

Both the last two are very hairy, the former all over, the latter especially on the nodes which look like little knees. Both like to grow on the woodland edge and hold the morning dew, and hence the mist, close to the ground.

Tufted Hair-grass
Deschampsia cespitosa (L.) Beauv.

A densely tufted grass with leaves like grass saws, for they are edged with hairs full of pure silica.

Bents
Agrostis spp.

The other components of good greensward.

Fine Bent
Agrostis tenuis Sibth.

Black Bent
Agrostis gigantea Roth

White Bent
Agrostis stolonifera L.

Two which thrive under meadow management:

Meadow Foxtail
Alopecurus pratensis L.

Marsh Foxtail
Alopecurus geniculatus L.

This prefers water meadows.

Sweet Vernal-grass
(See page 135)

Canary-grass
Phalaris arundinacea L.

This is like a small version of the Common Reed, and comes in a number of variegated varieties which may be planted round your garden pond.

CLASS 3: ORDER 3
PURSLANE FAMILY
PORTULACACEAE

Spring Beauty
Montia perfoliata (Donn ex Willd.) Howell

An introduction from the west coast of North America where it was discovered and used as a pot herb by the pioneers; it is still worth experimenting with. Our native near neighbour, found around mountain springs, is called Blinks. It got its name from the shy white flowers in hanging heads, a name which could never be applied to the American plant, for its wide-eyed flowers stare out from the cup-like sheath of two perfoliate stem leaves.

CLASS 4: ORDER 1
BUDDLEJA FAMILY
BUDDLEJACEAE

Buddleja
Buddleja davidii Franch.

Also known as the Butterfly Plant because of its attractive properties both to us and them. It was introduced around 1890 from China and now it is very much at home, a naturalised Londoner.

PLANTAIN FAMILY
PLANTAGINACEAE

This group of plants must be given pride of place in these Royal Lists, for they were the first true weeds to be put on record in the pollen sequences. No more need be said except perhaps that, even if they have lost their use in medicine, each year the Hoary Plantain produces an abundance of nectar for the insects and all three an abundance of seed for our song birds.

Rat's-tail Plantain
Plantago major L.

Hoary Plantain
Plantago media L.

Ribwort Plantain
Plantago lanceolata L.

Dogwood *(plate 15)*
Thelycrania sanguinea (L.) Fourr.

Dogwood, Cornel, Bloody Twig, Dog-berry, Hound's Tree, Gaten-tree—just a few of the common names given to this shrub-cum-tree which graces our limestone and chalk landscapes with its green and coral-red presence throughout the year. White flowers produced in June give rise to purple-black berries which yield one third their weight in oil, which can be used both in lamps and cooking.

Its wood, though small in size, is fine in grain, ideal for making toothpicks, lace-bobbins and butchers' skewers or 'dogs' for they impart no flavour to the meat. Its bark is bitter and astringent to the taste, and provides a good tonic draught, though not as good as its North American counterparts which also have much showier flowers.

VERBENA FAMILY   VERBENACEAE

Vervain
Verbena × hybrida Voss

A plant held in great esteem in many rituals and ceremonies, so much so that Pliny relates that the forefathers of Asterix 'collected it only when the great Dog Star was just rising in the heavens and, when plucked, an offering of honeycomb was to be made to the earth as a recompense for depriving her of so goodly a herb'. So highly was it valued in the past that even today as a weed it has its main stations in towns and around habitations. The Garden Verbena which can also become established as a weed, is usually a hybrid, one parent of which originated in South America.

BEDSTRAW FAMILY   RUBIACEAE

Members of this family may be known when not in flower because of the regular arrangement of their 'leaves' in rings. Here we must be careful of the term 'leaves', for a leaf by definition has a bud in its axil. If you look carefully you will see in each ring only two have such buds, the rest are not leaves but stipules, although they look the same and perform similar functions.

This family has provided the world with coffee, quinine and the garden Gardenia to name but one very beautiful plant.

Lady's Bedstraw
Galium verum L.

Laid between washed linen it sweetened the sheets of both ladies and gentlemen alike. In the field it provides its own special yellow-gold haze across the meadows.

Goosegrass
Galium aparine L.

Also known as Cleavers, Scratchweed, Catchweed and Sticky Willie, from its hooked hairs and habit of latching on to passers-by. It was long used, packed into a funnel, as a sieve for straining milk. It was also taken as tea or broth to keep those which would turn to fat both lank and lean, and the seeds roasted make a potable coffee.

CLASS 4: ORDER 2

Field Madder
Sherardia arvensis L.
(A member of the Bedstraw Family, see above)

Named after William and James Sherard, who befriended the German botanist Dillenius, described by Linnaeus as the 'foremost botanist of this age'. They brought him to England and he worked both in Oxford and Eltham in Kent, where he wrote his monumental *Hortus Elthamensis* illustrated with 417 drawings etched by himself. William Sherard died in 1728 and left his library and herbarium and a legacy to the University of Oxford—the latter to institute the Sherardian Chair of Botany, on condition that Dillenius should be the first professor.

FIGWORT FAMILY
SCROPHULARIACEAE (see also pages 133, 141 and 161)

Sharp-leaved Fluellen *(plate 5)*
Kickxia elatine (L.) Dum.

Its halberd-shaped leaves (which are bitter to the taste and were used to make an ointment for the skin) stood guard along the Gallery Bank of the Palace Gardens as did Shakespeare's Welshman at the battle of Agincourt.

Musk (Monkey Flower)
Mimulus moschatus Douglas ex Lindley

A native of the Pacific coast of North America it is cultivated both for its flowers and its musk-like smell. It is of interest that all the plants found in Britain today have lost their scent, perhaps through too much inbreeding.

ROSE FAMILY (see also page 150)

Parsley Piert
Aphanes arvensis L.

Lesser Parsley Piert
Aphanes microcarpa (Boiss & Reuter) Rothm.

Two halves of what is known as an aggregate species and very difficult to tell apart unless confirmed by experts. Both these small but troublesome weeds have their native home in this country: the former on base rich limestone and chalk around cliffs, the latter on more acid sandy heaths.

CLASS 4: ORDER 3

Fringed Pearlwort
Sagina ciliata Fries

Mossy Pearlwort
Sagina procumbens L.

Two members of the Campion Family (see also pages 149 and 150) these are mat-formers which can quickly take over an untended bed, moving in from the cracks in the path which they have already filled. Of all our weeds these have the smallest flowers, each one of which can be perfect although you will need a magnifying glass to seek and see that perfection for yourself. All the flowers of the former so far seen in the Palace Gardens are 4-merous, that is they have all their parts in fours. How about the ones in your garden? They may have theirs in fives.

CLASS 5: ORDER 1
IVY FAMILY   ARALIACEAE

The family which has provided the Eastern World with its 'universal remedy' Ginseng. Though its cure-all properties are doubted by most proponents of western medicine who relegate its powers to that of a 'universal placebo' it is apposite to remember the following facts: 80 per cent of the world's rural people today rely on traditional herbal medicines as their primary method of health care, contemporary world trade in medicinal plants tops £250 million per annum, and that at this moment

Meadow Vetchling
Lathyrus pratensis L.

Bird's-foot Trefoil
Lotus corniculatus L.

Greater Everlasting Pea
Lathyrus latifolius L.

(plate 41)

much international interest in homoeopathic and ethnic medicines has been rekindled, not just by the cranks but within the multi-national drug companies.

Ivy (plate 6)
Hedera helix L.

A fitting cloak for any Royal tree and note that it is cloak and not a shroud, for ivy does but little direct harm to its support. Ivy is an epiphyte, which means that it clings onto the outside gaining no direct sustenance from the tree. If it binds the trunk too tight distortion of both bark and wood may result, likewise its evergreen leafy canopy and spreading roots may compete with those of the tree. It may also help to make the tree top heavy and topple it in a high wind. There are however just as many instances where the ivy trunk itself props up the ailing tree.

The wood has been used for many things including the manufacture of the once famous ivy cups. Legend has it that wine drunk from cups made of ivy wood not only prevented the worst effects of hangover but cured ailments of the spleen. The Greek god Bacchus wore a wreath of ivy, and similar wreaths or ivy boughs were hung as advertisements outside establishments which sold wine.

The flowers of the ivy which open between November and February provide a surfeit of nectar for butterflies, bees, moths and many other winter-flying insects. Likewise its chocolate-black berries provide our overwintering songbirds with welcome fruit in these the harshest months of the year.

The most magic transformation is from the typical ivy-shaped leaves produced under dense shade to the almost undivided bright-green leaves of the mature flower-bearing shoots. Recorded in the Palace Gardens in the past it is no longer there, unless a bird has recently planted a seed ready potted for posterity.

PRIMROSE FAMILY   PRIMULACEAE

The family which gave us all our garden Primroses and Primulas, and our indoor Cyclamens.

Primrose (plate 7)
Primula vulgaris Hudson

Once common, so common even about London that every child could pick a bunch for Mum at Easter. Those days are alas gone; habitat destruction and greed of 'gardeners' has robbed so many of us of this very special part of our spring heritage.

Queen Victoria sent primroses to lay on Dizzy's (Disraeli's) grave with a card bearing the words 'His favourite flower'. They were indeed the favourite flower of the Prince Consort, not of Disraeli who preferred Dahlias.

Yellow Loosestrife (plate 8)
Lysimachia vulgaris L.

Believed to placate the tempers of horses and cattle let loose in the field, hence its name which ends (with) strife. It was also laid beneath the yoke of oxen for the same purpose. A handsome plant of the waterside and although of the same name, please note not of the same family, as its Purple neighbour.

Scarlet Pimpernel (plate 8)
Anagallis arvensis L.

Also known as the Countryman's Weather Wiser for its flowers shut tight before rain actually falls. Care must however be taken for even on a sunny day it shuts its flowers at around two o'clock in the afternoon.

The fruits are great fun for each is like a tiny ball with a schoolboy's peakless cap on top which opens to reveal itself brim-full of seeds. The same plant comes in four colour forms: white, pink, a gorgeous blue and scarlet red. The latter is by far the commonest and was held to have been stained by the blood of Christ. Sir William Hooker states that the name Anagallis comes from the Greek words 'again' and 'to adorn', for it comes up again and again to adorn our pathways and the edges of our fields.

PERIWINKLE FAMILY
APOCYNACEAE

Lesser Periwinkle (plate 9)
Vinca minor L.

This doubtfully native plant is a member of a mainly tropical family. Close inspection of the flower is well worth while, for it has a very special form which ensures cross-pollination—I will not spoil your own sense of discovery by describing it. However, when you have done this, keep a lookout for ripe fruits full of seeds; for though this is a very common plant, such fruits are very rare in Britain.

POTATO FAMILY   SOLANACEAE

A family which gave us the potato, tomato and tobacco, aubergine, paprika and both red and green peppers—all of which originated in South or Middle America—also provides our own hedgerows with some very poisonous plants.

Duke of Argyll's Tea-plant (plate 10)
Lycium halimifolium Miller

Introduced from south-east Europe around 1730, and long naturalised especially on walls. The Duke of Argyll received two plants wrongly labelled for his garden: one was a tea labelled 'Lycium', the other a Lycium labelled 'Tea'. The confusion has taken root in the common name of this once fashionable garden plant. The Duke was known rather denigratingly as the 'Treemonger' by Horace Walpole, so even in those days not all politicians understood the importance of plants.

Bittersweet (plate 10)
Solanum dulcamara L.

Compare the flowers with those on either your tomato or potato plants and you will see why they belong to the same family. You will also see that whereas your crop plants are all very similar to each other, the individuals of each species of our weeds show considerable variation. However, go no further with the bittersweet for this plant is poisonous. Indeed it gets its name from the fact that its roots warn you with a bitter taste, but leave the sweet taste—one presumes of melancholy if not death—upon your tongue.

Black Nightshade
Solanum nigrum L.

The ripe berries of this are black, not red like those of the last named. Though they are eaten with relish in other parts of the world, ours are poisonous, although some appear to be less so than others. This is all part of their natural variation. However as there is no quick safe way of telling, never be tempted to try.

Tomato
Lycopersicon esculentum Miller

Seedlings of this introduced plant were seen on the West Terrace in 1961. How they got there we will leave you to decide. When first introduced into Europe they were known as Love Apples, now they put sauce over more food than anyone should care to think about. The modern cultivars of this, and in fact of all our crop plants, are now so highly bred that they require constant attention including regular cross-fertilization with plants taken from the wild, thus topping-up the gene pool of the crop, strengthening and improving the plants and helping them to resist disease and environmental change.

This is the main reason why it is imperative that the world has large areas of its natural vegetation set aside as genetic reserves. It is of vital importance to all our futures.

Tobacco
Nicotiana sp.

If some garden-party goer can be blamed for any tomato seedlings, perhaps Prince Albert can be blamed for the presence of this plant. Prince Albert liked to smoke, Queen Victoria objected to the habit and she lived much longer than he. Thereby hangs a government Health and Wealth Warning.

FIGWORT FAMILY (see also pages 133, 138 and 161)

Common Mullein *(plate 5)*
Verbascum thapsus L.

A favourite with the bees and who can blame them, its 1.5m (5ft) tall spike of sulphur-yellow flowers being also called Aaron's Rod, High Taper, Candlewick Taper and many other names. Its flannel-like leaves remind us of those of the tobacco and of the affinities of this family with the last; they were used bound round the arms and legs to ward off the ague.

BELLBINE FAMILY
CONVOLVULACEAE

Lesser Bindweed *(plate 11)*
Convolvulus arvensis L.

This one climbs counter-clockwise, binding up the crops, and is thus a noxious weed. The only virtue it appears to have is the resin produced from its roots, though it is a poor substitute for scammony, a product of *Convolvulus scammonia* imported from the Levant. However, Culpeper says of the latter: 'I would advise my countrymen to let it alone, it will gnaw their bodies as fast as the Doctors gnaw their purses.'

Greater Bindweed *(plate 11)*
Calystegia sepium ssp. *sylvatica* (Kit.) Batt.

American Bellbine *(plate 12)*
Calystegia sepium ssp. *sepium* (L.)Br.

Both now thought to be subspecies, the former is native, the latter introduced not from America but from Europe. The green bracteoles which surround the base of the flower are inflated in the latter and, if squeezed sharply, the large showy flower pops out.

CLASS 5: ORDER 2
BELLBINE FAMILY (see also above)
Lesser Dodder
Cuscuta epithymum (L.) L.

A flower-bearing plant which lives as a total parasite on a number of plants both of the pea and of the heather family. That is should occur in London is strange, but it must have been brought in with the horticultural stock on which it was found growing. Culpeper states that it carries the virtues of the plant on which it is growing—in this case, solitude.

UMBRELLA FLOWER FAMILY
UMBELLIFERAE

A family which provides our roadsides and waste places with some of their most characteristic flowers and our cuisine with some of its most flavoured tastes—dill, caraway, fennel, parsley, carrots, celery, coriander, angelica and parsnip to name but a few, all of which come from such 'umble beginnings. Its members are usually readily recognised by the way their small flowers are arranged in large showy heads, each one at the end of what look like the spokes of a cartwheel or an umbrella. Hence the name Umbelliferae, for such an arrangement is called an umbel and, if its branches are branched, an umbel of umbels.

Golden Chervil *(plate 13)*
Chaerophyllum aureum L.

An introduction from southern Europe now naturalised especially in Scotland around Callander and Perth; there could thus be a Balmoral to the story.

Cow Parsley
Anthriscus sylvestris (L.) Hoffm.

Early flowering, its white cartwheels grace the hedgerows especially of southern England in April. Its finely divided leaves which give it another not so common name—

Nipplewort
Lapsana communis L.

(plate 42)

Lesser Goat's-beard
Tragopogon pratensis L.
ssp. *minor* (Mill.) Wahlenb

Sticky Groundsel
Senecio viscosus L.o.

Queen Anne's Lace—are favourite food of cattle, although its roots are poisonous to man. Like so many of our large weeds it denotes good ground, and has become commoner with the more widespread use of fertilizers which may drift from the fields into the surrounding countryside.

Knotted Parsley
Torilis nodosa (L.) Gaertner

An annual plant which often grows prostrate on the ground and produces small almost ball-like masses of pink flowers in June. Its outer fruits are very catchy numbers with spreading spines; the inner have little knots or tubercules, hence its name.

Pignut
Conopodium majus (Gouan) Loret

The nut-like tubers are good to eat, both cooked or raw. However, please be careful because this family includes a number of our most poisonous plants. In the past the tubers have been used in time of famine to produce flour, but have usually been fed to the pigs who seem to enjoy them just as much as truffles—a case of little pigs going down-market.

Hemlock Water Dropwort *(plate 13)*
Oenanthe crocata L.

Growing as it does close to water and especially around springs it could be mistaken for watercress, and has been mistaken for both celery and parsley. The mistake is usually not repeated, for the plant is poisonous. Anne Pratt records a case in which seventeen convicts working on the embankment near Woolwich dug up some roots and ate them with their dinner; all became ill and four died. There are many other similar cases on record, but always there is the problem of correct identification of the offending plant. Members of this family which contains Hemlock, Cowbane and many other poisonous plants are notoriously difficult to identify, especially when you are dead. Would-be amateur herbalists please note, you may do it to yourself but not to others for if your botany is at fault, 'quack quack' can become 'croak croak'.

Fool's Parsley
Aethusa cynapium L.

Another poisonous plant not usually mistaken for real parsley for it has an unpleasant odour and darker leaves. However beware, especially where children are concerned.

Hogweed
Heracleum sphondylium L.

Heracleum from the sword of Hercules; Hogweed because pigs, horses and especially cattle dote on it. It was at one time collected in Britain and used to try to make sugar, an attempt which failed because of the amount required. In Asia it has been employed as a narcotic in certain 'religious' ceremonies, and is said to inflict the users with a violent desire for self-destruction.

It is of great interest that many plants which have a certain chemical property in one area, have no such property when growing or even grown in another.

Giant Hogweed *(plate 13)*
Heracleum mantegazzianum Sommier & Levier

Like the last plant but a truly worthy sword of Hercules, for it grows up to 4m (14ft) tall. It might also be called the Sword of Damocles for real trouble hangs over the head of anyone who is allergic to its sap. I once saw two students who had made pea-shooters out of its hollow stem; it was impossible to recognise them for their faces and hands were red and swollen out of all recognition. Introduced from the Caucasus, it is now running wild through many of our lowland river valleys. The best advice is that, unless you are an expert, enjoy the members of this family from a discreet distance.

MARSH PENNYWORT FAMILY HYDROCOTYLACEAE

A family consisting of but one genus and seventy species.

Marsh Pennywort
Hydrocotyle vulgaris L.

Pennywort because of the shape of its leaves and Marsh because of its preferred habitat. Of all the plants in this treasury, these may well be direct descendants of the wild stock which grew naturally in the marshy ground watered by Tyburn. If so, their perennial presence may still contain molecules from both noblemen and commoners who suffered a fate worse than death in that infamous place. Its other common name, White Rot, comes from its white flowers dominating areas in which sheep suffered from liver fluke. The only thing which the flukes and the plant have in common is their need for marshy ground in which to complete their life cycles.

ELM FAMILY ULMACEAE
Wych Elm
Ulmus glabra Hudson

English Elm
Ulmus procera Salisb.

Cornish Elm *(plate 14)*
Ulmus minor var. *cornubiensis* (Weston) Rehd.

All three are present in the Palace Gardens, all having been planted. The first has very short leaf stalks, each less than 3mm ($\frac{1}{10}$in) long.

The word 'wych' has nothing to do with the black arts but comes from the Anglo Saxon *wichen* meaning to bend, its flexible small branches being woven into wattle.

The English elm has rough dark-green leaves which sting some people like nettles, while the third, perhaps planted to make the Duke of Cornwall feel at home, has smooth light-green leaves. Each tree is a veritable high-rise hotel, providing shelter and food for a whole host of insects. It was an elm tree under which the Duke of

Wellington stood at the Battle of Waterloo. You too may have a similar experience if you stand beneath the truly gigantic Waterloo Vase near the Rose Garden and review the ordered mêlée of a garden party. Our elms are today fighting their own battle against the Dutch Elm disease and are, unfortunately, losing.

The English elm is not only native but may well be endemic, meaning that it does not grow naturally anywhere else. It would be a sad day for England if it became extinct. Work is however underway in a number of places to breed an English elm which is resistant to the disease, and in the meantime an American hybrid elm, Sapporo Autumn Gold, is being planted to cover the gap so that we don't lose the insects before we can replace the trees.

Elm wood is very durable and second only to oak for the construction of boats. Hollowed-out trunks were used as water-pipes, and examples still in working order are excavated to this day from beneath the streets of London.

GOOSEFOOT FAMILY CHENOPODIACEAE

All-seed
Chenopodium polyspermum L.

Fat Hen
Chenopodium album L.

Fig-leaved Goosefoot
Chenopodium ficifolium Sm.

Red Goosefoot
Chenopodium rubrum L.

Goosefoots, named from the shape of their leaves, are common weeds originating probably from the open refugia along the sea coast and moving inland as the forests were cleared for agriculture.

All-seed is aptly named for its many brown shining seeds take the place of the less conspicuous flowers in late summer. Fat Hen derives from the well-fed appearance of the green-white flower masses, and perhaps also from its abundance in farmyards where the hens gorge themselves on both its leaves and flowers. Fig-leaved speaks for itself, and this plant thrives around manure heaps; while the red-brown seed masses give red goosefoot its name.

It is a pity that *Chenopodium bonus-henricus*, Good King Henry, does not grow in the Palace Gardens, for its leaves when boiled are good to eat. I have, however, seen it growing in profusion at Hampton Court.

Goosefoots are good for the garden; their roots penetrate deep bringing up the nutrients, and composting of their remains returns the latter to the surface ready for use.

CLASS 5: ORDER 3
ELDER FAMILY   CAPRIFOLIACEAE

Elder *(plate 15)*
Sambucus nigra L.

Great plates of creamy blossom, great masses of sickly fruits followed by bottles of refreshing wine, best made

from the flowers. It is a plant which likes the company of man, delighting in best-forgotten places which are enriched with nitrogen and phosphorus. It is also a sad tree, the reason perhaps best stated in this poem by P. J. Kavanagh:

Feigns dead in winter, none lives better.
Chewed by cattle springs up stronger; an odd
Personal smell and unlovable skin;
Straight shoots like organ pipes in cigarette paper.
No nurseryman would sell you an
Elder—'not bush, not tree, not bad, not good'.
Judas was surely a fragile man
To hang himself from this—'God's stinking tree'.

In summer juggles flower-plates in air,
Creamy as cumulus, and berries, each a weasel's eye,
Of light. Pretends it's unburnable
(Who burns it sees the Devil), cringes, hides a soul
Of cream plates, purple fruits in a rattle
Of bones. A good example.

(From *Life before Death* (Chatto & Windus with Hogarth Press, 1979)

CLASS 5: ORDER 5
FLAX FAMILY   LINACEAE

Linseed (Flax)
Linum usitatissimum L.

A plant of inestimable value, but its origins even today remain obscure. Cultivated for its fibres which are spun to make the finest linen, and for its seeds which are crushed to obtain oil of linseed, it has been grown throughout the civilised world for many centuries. Its native home is not known.

Purging Flax
Linum catharticum L.

Another name for this delicate little plant is Fairy Flax. It thrives on open chalky soil and is often common on hillsides hence its other name, Mill Mountain. *Catharticum* means 'purging', and it is. It was also recommended in the past for rheumatism.

CLASS 5: ORDER 6
HORSE CHESTNUT FAMILY HIPPOCASTANACEAE

Horse Chestnut
Aesculus hippocastanum L.

If it wasn't for this introduction from Albania or Greece, the annual ever-so English game of conkers would never have come into being. Likewise if it wasn't for this pleasant pastime, the tree would probably be much commoner than it is, and those trees which have got past the seedling stage and reached maturity would probably be in much better shape.

There are two magnificent trees in the Palace Gardens: one planted by HRH Prince Albert in 1914 and another planted by HRH Prince George in 1917. They are presumably the source of the conkers from which the seedlings found in the garden grew.

CLASS 6: ORDER 1
LILY FAMILY   LILIACEAE
Lily of the Valley *(plate 16)*
Convallaria majalis L.

A native plant of dry woodlands on lime-rich soils. Grown in gardens (but please never transplanted from the wild) it can become a real weed.

Held in great esteem in the past as a cure for gout, loss of memory and apoplexy, its flowers dried and taken as snuff or steeped in wine were said to alleviate headaches. The flowers have been used in the manufacture of certain perfumes.

Fritillary (Snake's Head) *(plate 17)*
Fritillaria meleagris L.

One thousand of these handsome flowers were planted in 1963 and some have become naturalised in the Palace Gardens.

There once was a meadow between Mortlake and Kew named Snake's Head Meadow, because of the abundance there of this flower. A typical plant of water meadows along the Thames and other rivers, now much lost thanks to drainage.

Bluebell *(plate 18)*
Endymion nonscriptus (L.) Garcke

In the rambling and bicycling 1920s and 1930s, enormous quantities of these plants were culled from our coppice woodlands around London. If only they had been picked rather than pulled, which damages the white bulb, it would have done little harm. I wish it could be so again; perhaps one day it will.

Grape Hyacinth *(plate 16)*
Muscaria spp.

Grape Hyacinth, is it a weed? Please come and look at my rose bed, come on in and dig it out, year after year after year. *Muscari atlanticum* Boiss. & Reuter, is a native plant which was introduced into gardening and has gone wild.

DAFFODIL FAMILY AMARYLLIDACEAE
Spring Snowflake
Leucojum vernum L.

Possibly native, but a very rare plant in the wild. Twenty-five were planted in the Palace Gardens in 1963.

Snowdrop *(plate 19)*
Galanthus nivalis L.

Six thousand were planted in 1960 and two thousand in 1961, and this probably native plant is doing well.

Pheasant's-eye Daffodil
Narcissus majalis Curtis

Primrose Peerless
Narcissus biflorus Curtis

These, along with other cultivated varieties, have in recent years been planted in great number and all appear to be well-settled to Palace Gardens' life.

The bulb of the latter—and of certain daffodils—when dried, provides a flour which is such a strong emetic that it has been given the name *Bulbi vomitarii*. An extract was also used for whooping cough, one presumes with spectacular results.

RUSH FAMILY   JUNCACEAE

Before carpets were introduced into the richest houses, like that of Edward III, in the thirteenth century, rushes of various sorts were strewn upon the floors. They were also used as wicks for candles. It is reported that Thomas à Becket 'was manfull in his household, for his hall was everie daye, in somer season, strewed with greene rushes . . . for to save the knyghtes' clothes that sate on the floor for defaute of place to sit on'.

Slender Rush
Juncus tenuis Willd.

Toad Rush
Juncus bufonius L.

Hard Rush
Juncus inflexus L.

Soft Rush
Juncus effusus L.

Jointed Rush
Juncus articulatus L.

Please don't muddle them up with the grasses. Most have round stems and stem-like bracts. However a close look at their flowers will reveal tiny 'lilies'.

Field Wood-rush
Luzula campestris (L.) DC.

Wood-rushes are distinguished from those above by the possession of very long white to colourless hairs which fringe the leaves, especially at the base.

CLASS 6: ORDER 3
Docks
Rumex spp.

Members of the Knotgrass Family (see also page 148), the docks differ from knotgrasses by having their flowers arranged on many long branches.

Sheep's Sorrel *(plate 20)*
Rumex acetosella L.

The hastate (spear-shaped) leaves are acid to the taste, and make a pleasant addition to salads.

Common Sorrel
Rumex acetosa L.

In this species the upper leaves clasp the stem, hence its meaning 'parental affection' in the Language of Flowers. The leaves of this species may be used in salads and sauces, and the roots have been used in times of famine to make flour and bread. The roots yield a good red dye.

Great Water Dock *(plate 20)*
Rumex hydrolapathum Hudson

The largest of our docks, it grows by the waterside and its roots may be used to make an astringent mouthwash.

It is suggested that this, or one of those below, is the Herba Brittanica of Pliny and is derived, not from the word Britain but from Brit, to tighten, Tan a tooth, and Ica loose. It is a useful remedy for tightening loose teeth, and the powdered root makes an excellent dentifrice, but it must be used fresh.

Curled Dock *(plate 20)*
Rumex crispus L.

The wavy edges to its leaves give this one away. Its roots have been used to treat conditions of the skin.

Broad Dock
Rumex obtusifolius L.

Every child knows that the sting of a nettle may be assuaged by the application of the juice from a dock leaf. Without doubt this is the best example of the old belief that a poisonous plant and its antidote invariably grow together.

Wood Dock
Rumex sanguineus L.

In its normal common form it has green veins to its leaves. However red-veined forms were found and thought to be so efficacious for the liver, and hence the blood, that they were cultivated and became commoner. Mixed with sorrel and dressed as spinach they make a tasty dish.

Clustered Dock
Rumex conglomeratus Murray

Though not the Fiddle Dock, the leaves of this are often panduriform or fiddle shaped. Of all docks this has the sharpest, most acid taste and is not recommended for the salad.

CLASS 6: ORDER 5
WATER PLANTAIN FAMILY ALISMATACEAE
Water Plantain
Alisma plantago-aquatica L.

The bulbous roots of this handsome water plant yield an edible flour; it was also said to cure hydrophobia and anyone who had eaten a 'sea hare'. As only someone suffering from some delirium would think of eating a sea hare—a very weird sort of marine 'slug'—there could be some connection, but probably no cure.

CLASS 8: ORDER 1
MAPLE FAMILY   ACERACEAE
Sycamore
Acer pseudoplatanus L.

Maple is today such a common tree, and perhaps the one which most deserves the name of a weed thanks to its autogyro fruits, that it may seem difficult to believe that it is not a native. It was probably introduced in the fifteenth century, and was thought to be the tree into which Zaccheus climbed in order to see Christ. At one time a fashionable tree to plant on great estates, it is now at home even in high moorland farmyards where it thrives on dung and, though shaped by the wind, is never bent. Grown at one time for its clean white wood which, being tolerant of soaking and redrying many times, was used to make the rollers in textile mills and home mangles. Today it is of great value as a veneer, but only for the inside of cabinets and the like for it discolours in the sun.

Related to the North American Sugar Maple, it may be tapped in spring and yields a wort for making ale with a great saving of malt.

HEATHER FAMILY (see also page 149)
ERICACEAE
Heather *(plate 21)*
Calluna vulgaris (L.) Hull

The plant which more than any other graces our moorlands with its presence and feeds both bees and grouse. Those of you who enjoy the lamb culled from those same moors, think when you decry the Glorious Twelfth and everything it stands for. Even if you only eat heather honey, remember that without constant management the heather crop on those same moors would soon decrease. The management of our countryside is a very complex business, for it provides a livelihood for so much wildlife and for so many people.

Cross-leaved Heath *(plate 21)*
Erica tetralix L.

Also known as Bog Heather and Honey Bottle it likes to grow in wetter places than heather and, though visited by bees, Gerard said 'of these flowers bees do gather bad honey'.

Dorset Heath *(plate 21)*
Erica ciliaris L.

A native plant of Dorset, south Devon and west Cornwall. The word 'heath' means solitude, and it was from the solitude of such poor land that many of our choicest garden plants have been derived. Of all our British vegetation, our lowland heathlands have suffered most at the hands of agriculture and of fire brought on by visitor pressure. In many places the heaths have all but disappeared and with them have gone not only their characteristic plants and animals, but also the solitude still sought by many people.

(plate 43)

Smooth Sow-thistle
Sonchus oleraceus L.

Corn Sow-thistle
Sonchus arvensis L.

Gallant Soldier
Galinsoga parviflora Cav.

Oxford Ragwort
Senecio squalidus L.

WILLOW-HERB FAMILY
ONAGRACEAE
Willow-herbs
Epilobium

The young shoots of many of the species, especially the larger, have been collected and dressed as a palatable substitute for asparagus.

Great Willow-herb (Codlins and Cream)
(plate 22)
Epilobium hirsutum L.

Great it is, our largest growing to 150cm (60in); and there is no disputing its hirsute character. Codlins from the scent of the flowers which is like cooked fruit; the white stamens and stigma putting the dob of cream on top.

Broad-leaved Willow-herb
Epilobium montanum L.

Not only of the mountains but certainly broad-leaved and a common garden weed.

Spear-leaved Willow-herb
Epilobium lanceolatum Seb. & Mauri

A plant of roadsides, railway banks, walls and dry waste places, one wonders where it originated from, although its occurence along the coast and by our estuaries in the south perhaps gives the clue.

Pale Willow-herb *(plate 22)*
Epilobium roseum Schreber

A plant of wet places where it certainly must have survived throughout the long period of forest cover to emerge to take up the ruderal habit, once man had removed much of the forest.

American Willow-herb
Epilobium adenocaulon Hausskn.

A native of North America it was first recorded in Britain in 1891, since when it has spread across the land and is now found growing on both the Inner and Outer Hebrides.

Short-fruited Willow-herb
Epilobium obscurum Schreber

Like the last, liking damp places where it spreads by means of short stolons which are produced only in late summer.

Rosebay Willow-herb *(plate 22)*
Chamaenerion angustifolium (L.) Scop.

Differing from all the above in that it has all its leaves arranged in a spiral around the stem, none are opposite each other. This is the plant which took over the bombed sites during World War II. Though now looked upon as a native plant, at one time it was thought to have been introduced. Indeed Gerard speaks of it as only growing in one locality in Yorkshire from whence he obtained it for his garden.

CLASS 8: ORDER 3

KNOTGRASS FAMILY
POLYGONACEAE

This is a family of notorious weeds, although if put in the dock and tried for its virtues it would be found to yield some very useful garden plants including rhubarb and buckwheat. We know the latter shed its pollen over Londinium, and Gerard records its cultivation in England in 1597. Cakes and hasty puddings made from its flour are very tasty, and the crop may well come to the aid of the increasing number of people who are developing allergy to farina made from members of the Grass Flower Family which includes all the cereals. In the *Biblia Dudesch* printed in Germany in 1552, the translator renders Isiah 28: 25 as 'he soweth buckwheat'.

Common Knotgrass *(plate 23)*
Polygonum aviculare L.

Growing up to 2m (6ft) tall, it produces a great abundance of seed of great importance to wild birds. Its name comes from the notion that a decoction arrested the growth of both children and sheep. Shakespeare knew this for we find in *A Midsummer Night's Dream* 'You minimus of hindering Knot Grass made'. No mention is found of this in Culpeper who states:

. . . helps spitting and pissing of Blood, stops the terms and all other fluxes of blood, vomiting of blood, weakness of the Back and Joynts, Inflammations of the Privitus, and such as piss by drops, and it is an excellent remedy for Hogs that will not eat their meat. Your only way is to boyl it in its prime about the latter end of July or beginning of August.

The long list infers that such a common plant was used in an attempt to treat many illnesses and the last point is not so much a let-out as an acknowledgement of the fact that even the proven properties of plants change through the seasons.

One fascinating record concerning the use of this plant is as food for silkworms. It is stated in the *Diario Mercantile of Venice* that in 1852, silkworms were raised and silk produced from them in sixteen days when fed on this plant, even in winter. If only James I had known.

This is an aggregate species and includes also

Small-leaved Knotgrass
Polygonum arenastrum Boreau

Amphibious Persicaria *(plate 24)*
Polygonum amphibium L.

The latter is a floating plant of the edges of ponds and lakes and a noxious long-term weed of drained land, so perhaps again those in the Palace Gardens are direct descendants of the ones which grew in the waters of Tyburn. The roots were used and preferred in Lorraine as a substitute for sarsaparilla.

Red Leg
Polygonum persicaria L.

The name comes from its red stems which are swollen at the nodes to form knees the like of which any Queen's Scout would be justly proud.

Pale Persicaria
Polygonum lapathifolium L.

Similar to the last but with greenish-white rather than pink flowers. Persicaria means 'restoration' in the Language of Flowers.

Black Bindweed *(plate 23)*
Bilderdykia convolvulus (L.) Dum.

A really nasty customer in wheatfields where it takes over, twisting round and binding the crop together. Finally its weight brings the wheat stalk to the ground.

Japanese Knotweed *(plate 24)*
Polygonum cuspidatum Siebold & Zucc.

Lesser Knotweed
Polygonum campanulatum Hooker

Giant Knotweed
Polygonum sachalinense F. Schmidt petrop.

Three introductions, the first from Japan, the second from Himalaya and the latter from the island of Sakhalin. All like to grow near water but once they have a roothold, watch out! They are all, however, good for the insects, their sweet-scented flowers being their great attraction. The third named is a real giant and grows to 4.5m (15ft) tall with leaves the size of handkerchiefs.

During Childermas, which recalls the Massacre of the Innocents, every child was wont to be whipped with sufficient gentleness to mark in his or her memory the day's events. The knotweeds provided rods for these ritual whippings, earning them the name of Discipline Scourge and Holy Innocents' Wort. The rods of these large introduced species are today children's playthings and, being hollow, break on contact causing little discomfort.

CLASS 10: ORDER 1
HEATHER FAMILY (see also page 146)

A good diagnostic feature of the members of this family is that, instead of bursting their pollen sacs, they release the pollen through special pores.

Strawberry Tree
Arbutus unedo L.

A native of the warm, wet, west coast of Ireland. Grown in gardens for its pink-tinged creamy white flowers and red fruit, and if you have the luck of the Irish it may become naturalised.

Rhododendron
Rhododendron spp.

Rhododendron ponticum L. was introduced from Spain and has now run riot over estates and large gardens where it causes immense harm by precluding woodland regener-ation and the growth of any ground cover. The Palace Gardens boast no less than a hundred cultivars which had their origins in Eurasia and North America. Fortunately they are all kept well in hand.

CLASS 10: ORDER 2
SAXIFRAGE FAMILY
SAXIFRAGACEAE
Strawberry Saxifrage
Saxifraga stolonifera Meerb.

A garden plant which, thanks to its stolons, can run wild and steal more space than it was given.

CLASS 10: ORDER 3
CAMPION FAMILY
CARYOPHYLLACEAE
White Campion
Silene alba (Miller) E. H. L. Krause

A short-lived perennial, annual or biennial plant and hence a promising weed. In most cases its flowers are only of one sex but infestation of the female flowers by the smut fungus causes them to produce stamens, the anthers of which are filled with violet fungal spores in place of pollen grains. In nature it hybridises with the Red Campion to produce great drifts of intermediates.

Common Chickweed
Stellaria media (L.) Vill.

Its three styles give this one away and despite its small size and annual habit, once established in your garden you will wish that you could do just that. It also has purple stamens, three if growing in deep shade and up to eight if growing in full sunlight.

The secret of its success lies at least in part in the abundant seeds. An average plant goes on flowering throughout the year producing 7,000 seeds, and with three potential generations a year that makes a possible 333 thousand million seeds from one plant. Get digging or keep chickens, for this humble weed once went under the name of Hens' Inheritance for where the gardener didn't do his job to perfection, the hens cashed in with fresh salad and seeds every day of the year. No wonder it is known as chickweed.

Lesser Stitchwort
Stellaria graminea L.

How did the stitchworts get their common name? Perhaps from the way they creep across the ground, rooting or sending up flowers as they go like a row of stitches. It is smaller than its larger brother the Greater Stitchwort which flowers earlier in the year. Both however are very beautiful plants and it seems strange that plant breeders have never managed to tame their somewhat straggly appearance and turn them into garden flowers.

CLASS 10: ORDER 4
CAMPION FAMILY (see also above)

Common Mouse-ear Chickweed
Cerastium holosteoides Fries

This plant is found growing at an altitude of almost 1,219m (4,000ft) in Scotland, and as no part of Britain goes much higher it may be looked upon as our most elevated weed. Its four, five or six styles will usually serve to distinguish it from the very similar chickweeds. Stoop down and take a look at this little hairy plant which creeps prostrate across the ground, only its flowering shoots lifting up against the force of gravity to bear its seeds of success as high as it can.

Corn Spurrey
Spergula arvensis L.

Although a troublesome weed on acid lands, some larger forms are grown as fodder plants and have proved to be very productive under adverse conditions.

SORREL FAMILY   OXALIDACEAE
Sleeping Beauty
Oxalis corniculata L.

Small Pink Oxalis
Oxalis corymbosa DC.

Not unlike geraniums at first glance but the fruit has no beak, and the leaves are made up of separate leaflets, often three in number, which show sleep movements folding down at night.

Sleeping Beauty with its yellow flowers was first recorded wild in Britain in 1585 but was introduced from abroad—where exactly is not known. It is of great interest that a very similar but smaller flower is an abundant native of New Zealand and Tasmania. The small pink oxalis, though a native of South America and though its fruits are unknown, is naturalised in many London gardens. It reproduces by means of bulbils.

The leaves of this genus, and especially of the Wood Sorrel our commonest species, are good to eat when young being rich in Vitamin C. As they get older they become charged with oxalic acid and may, like those of rhubarb, be used for cleaning brass, and not for eating.

CLASS 11: ORDER 1
LOOSESTRIFE FAMILY LYTHRACEAE

Purple Loosestrife
Lythrum salicaria L.

A waterside plant of great beauty, being both common and abundant in the south until drainage of so many of our wetlands robbed the country visitor of so much. Its willow-like leaves are full of tannin and it is reported to have been used in the past in the preparation of leather.

CLASS 12
ROSE FAMILY   ROSACEAE
'By any other name would smell as sweet' (see page 138).

CLASS 12: ORDER 1
Plums, Blackthorn and Cherries
Prunus

Wild Plum *(plate 25)*
Prunus domestica L.

Blackthorn (Sloe) *(plate 25)*
Prunus spinosa L.

The name wild plum is erroneous for nowhere is it truly wild or, if it is, we don't know where. *Domestica* is much better, for that is its origin probably as a hybrid between our native sloe and the cherry plum which was introduced from Asia or Europe. Selection and hybridisation have provided our modern orchards with a great variety of such fruits, the most famous being named after Queen Victoria, who viewed this Palace Garden from her special seat which is still there, set about a Plane tree.

Blackthorn, as its name implies, is well-armed compared with its domestic descendant.

Wild Cherry or Gean *plate 27)*
Prunus avium (L.) L.

Bird Cherry *(plate 26)*
Prunus padus L.

Cherry-laurel *(plate 26)*
Prunus laurocerasus L.

It was from the first wild native stock that selection and plant breeding gave us our so sweet cherries. The second, also a native, is common in the north; but is planted in the south for its beautiful though ephemeral masses of flowers.

The third was introduced from southern Europe and is an evergreen shrub which is second only to the Rhododendron in providing cover and unfortunately dense shade. Its erect flowerheads, which are not unlike those of the Horse Chestnut though smaller, decorate its dark-green foliage in early summer.

CLASS 12: ORDER 2
Hawthorn or May *(plate 28)*
Crataegus monogyna Jacq.

May flowers have been used as protection for the home and its inhabitants since at least Roman times. In England it formed the basis for many country ceremonies and feasts which welcomed summer in, and strict instructions were given for its arrangement around the windows. First to the east in honour of the rising sun, next to the south in honour of mankind, then to the north to procure protection for the family, finally to the west as a tribute to the Lord of Death. In these days of hyper-marts and deep freezes it is difficult for us to understand the real significance of rejoicing that the bad times of winter had gone and summer was just around the corner.

The superstition which still remains about bringing May flowers into the house is easy to understand. Though they may smell sweet in the field, confined in the house

The Queen's Own Hawkweed
Hieracium lepidulum (Stenström) Omang
var. *haematophyllum* Dahlst.

(plate 44)

and dying they reveal their true scent which is like rotting fish.

The most famous hawthorn is that planted by mistake by Joseph of Arimathea when he first came to Britain to preach the gospel. He landed on the Island of Avalon in what are now called the Somerset Levels and there where his staff broke into bud and bloomed he built a church which later became Glastonbury Abbey. There his thorn, or at least its descendants, still flower both in May and at Christmastide.

It is said that Puritans who attempted to cut down the tree failed, for their eyes were put out by the thorns or else they became lame. One can only wish that some equal retribution might fall on those who continue to destroy the natural wonders of those same Levels.

CLASS 12: ORDER 3

Rowan *(plate 28)*
Sorbus aucuparia L.

Whitebeam *(plate 28)*
Sorbus intermedia (Ehrh.) Pers.

Two members of another complex group of plants whose correct identification requires expert knowledge or help. The first is the easiest to determine and the most constant in its characters, its leaves being made up of more than four pairs of leaflets. The fruits, which have only one or two seeds and therefore should not be called berries, are well known and well-loved as rowan jelly. The other is one member of another aggravating apomictic (at times producing viable seed without fertilization) aggregate group whose white, almost furry leaves give all of them their common names.

CLASS 12: ORDER 4

Meadowsweet *(plate 29)*
Filipendula ulmaria (L.) Maxim.

'The leaves and floures farre excell all other strowing herbes, for to decke up houses, to straw in chambers, halls and banqueting houses in the sommer time; for the smell thereof makes the heart merrie, delighteth the senses.' So wrote Gerard of this plant in his Greate Herball of 1633, and so it was used in the days before carpets, and table manners.

The older generic name *Spirea* comes from Spireon or Garland Flower; although as its flowers fade rapidly once picked, it would not appear to be the best choice. However the scent—a sweet mixture of new-mown hay and soap—lingers on and can become quite nauseous. Linnaeus records that its roots which are a favourite food of pigs may be dried, ground, and provide a not bad substitute for flour. And beer has been made from its flowers.

Crab Apple *(plate 30)*
Malus sylvestris Miller

An undershrub especially of oakwoods and, though native many found today in our woods and hedgerows have probably reverted from cultivated forms. As apple pips don't breed true, all productive varieties are propagated by grafting originally onto wild stock, such stock providing as it were a background of wild strength to nurture the more highly bred scions.

CLASS 12: ORDER 5

Silverweed
Potentilla anserina L.

Midsummer Silver, Silver Fern, Fern Buttercup are just some of its other common names, but perhaps most apt in the Palace Gardens is Princes Feathers for the silver compound leaves might well adorn the device of the Prince of Wales. Traveller's Ease is yet another name which may well come from its beauty as a wayside plant or from the fact that it travels with ease all by itself by means of runners, and will soon be the ruination of your lawn.

The leaves retain their beauty when dried and are of great use in dried-flower pictures (but please be careful what other plants you pick), and the roots have a sweet taste especially when roasted. They have in fact helped to feed crofters in times of scarcity on the islands of Coll and Tiree where the plant grows abundantly, binding together the blown sand.

Creeping Cinquefoil
Potentilla × *italica* Lehm.

Cinq = five, foil = a leaf; *Potentilla* perhaps for its supposed potency as a medicinal herb though, as already stated, most of our commonest plants were attributed with such powers before the days of modern drugs. However we do know that one of this plant's parents was used as a febrifuge to bring down the temperature of the patient. The X again signifies a hybrid which indeed it is—a hybrid between two of our commonest plants, the common tormentil and the creeping cinquefoil—neither of which have been recorded in the Palace Gardens.

Dog Rose *(plate 31)*
Rosa canina L.

Greek fable holds that the rose was thornless until one day Cupid, stooping to pluck a newly opened bud, was stung angrily on the lip by a bee. Weeping, he ran to his mother who, to pacify him, strung his bow with bees, first plucking from them their stings which she placed upon the stem of the offending rose.

The rose in all probability because of its beauty and perhaps because of its armour, was used from early times as an heraldic symbol. It emblazoned the shields of Roman warriors, and Charlemagne carried roses on a golden field. At what date it became the distinctive badge of England is still unknown. Edward I adopted it as his cognizance, but long before that the Saxons carried a banner wrought with roses into the battlefield. The first appearance of the flower on the coins of our realm was the 'rose noble' in the reign of Henry VI, but it does not appear on the Great Seal of England until the reign of Edward IV.

Our garden roses are hybrids of complex origin and our wild roses and their natural hybrids are just as complex.

(plate 45)

Dandelion
Taraxacum officinale Weber

Slender Hardhead
Centaurea nigra ssp. *nemoralis* (Jord.) Gugl.

Spear Thistle
Cirsium vulgare (Savi) Ten.

Lesser Burdock
Arctium minus Bernh.

(plate 46)

Cut-leaved Bramble *(plate 29)*
Rubus laciniatus Willd.

Another Bramble
Rubus rubritinctus W.C.R. Watson

Willd. is the abbreviation for the name of the botanist Willdenow, but 'wild' as an epithet aptly describes the brambles in all their prickly arching glory, and the gory legs resulting from any explorations in the bramble patch. The botanist is also unlikely to emerge unscathed if he or she tries to put specific names to the ever so common brambles, because they are amongst our most difficult plants to identify.

Some floras lump them all together under *Rubus fruticosus* agg. The 'agg' could well stand for the aggro caused when trying to identify them, but in actual fact stands for 'aggregate species' which means much the same—try to identify them, if you dare. If you do dare you will need a copy of W. C. R. Watson's *Handbook of the Rubi of Great Britain and Ireland* (1958), which gives detailed descriptions of no less than 386 species, all of which have been found there. The cut-leaved bramble is but one of them, and a strange one, for it is unknown as a wild plant despite the name of its authority.

The problem is that brambles are apomictic, which means that they can at times produce viable seed without fertilization, this virgin birth of course resulting in progeny which look exactly the same as mum. However, every now and again true fertilization will take place and a new and distinct line may be set off on its apomictic way. Hence all the supposed different species, the identification of which is elementary—but only for Mr Watson.

To confuse the issue further, we also call bramble fruits 'blackberries', which they certainly are not. A berry by definition contains many seeds; but each of the fruitlets which make up the bramble contain only one hard seed (as those with false teeth will know to their cost) and should therefore be called drupelets. The whole is thus an aggregate or ataerio of drupelets. Confusing, but delicious when cooked with apples and as jelly or jam, and well worth all those scratches.

CLASS 13: ORDER 1
WATER-LILY FAMILY
NYMPHACEAE
White Water-lily *(plate 32)*
Nymphaea alba L.

It is said that the flowers open at 7am and close at 6pm; perhaps any of you with Greenwich Meantime to spare might check it out. A plant of slow rivers, ox-bows and still ponds, its alabaster blossom must be regarded as the queen of any watercourse. To call it a weed is surely sacrilege, but any water bailiff will tell you of the choking problems once it has got roothold in a reach or pool. The rhizome and roots are still used as natural dyes mordanted grey, chestnut or even dark brown, while Culpeper states: 'Water lilies are cold and dry, and stop lust. I have never dived so deep to find any other virtue the Roots have.'

LIME FAMILY   TILIACEAE
Lime (Linden) *(plate 33)*
Tilia × *europaea* L.

A hybrid between the native Large-leaved and Small-leaved limes. Doubtfully native, it is widely planted and I believe it was the pollen from such a tree which made the guardsman, of Chapter 2, sneeze.

CLASS 13: ORDER 5
BUTTERCUP FAMILY (see also next Order)
Columbine *(plate 34)*
Aquilegia vulgaris L.

'Columbine' refers to the shape of the flower which is said to resemble a nest of doves; but *Aquila* (eagle) enters this tranquil picture for the hooked spurs look not unlike the beak or talons of this bird of prey. Combined with a red Rose it forms the heraldic badge of the Royal House of Lancaster.

Gerard records that this is to some '*Herba leonis*, or the herbe wherein the Lion doth delight'; and Culpeper suggested that if it did it might cure its throat made sore perhaps by too much roaring.

In strict botanical terms the fruit is a follicle and in the Language of Flowers it means 'folly'. I wonder which came first.

CLASS 13: ORDER 7
BUTTERCUP FAMILY
RANUNCULACEAE
Traveller's Joy *(plate 35)*
Clematis vitalba L.

Along with Honeysuckle this is our only native woody climber; the nearest thing we have to a tropical Liana, it dangles from the trees in the chalk and limestone districts of the south, tempting every budding Tarzan to risk both life and limb.

Like some of the many showy garden Clemati, its fresh leaves are corrosive to the skin causing nasty lesions. So take care, unless you are courting sympathy, as it is said was the wont of beggars who used these as blister leaves in times gone past. When dried, the poison disappears and the leaves provide good fodder for cattle.

The young hollow stems of Clematis have been used as a form of self-consuming smoking pipe, the smoke evidently giving some pleasure complete with an acrid health warning. Where it grows in abundance its strong but pliable woody stems were used in the manufacture of baskets, bee-skips and even hurdles.

As Old Man's Beard it livens up the autumn scene with its masses of silk-tailed fruits. Its name Traveller's Joy, first coined probably at the time of Gerard, came perhaps from the fact that it is common along the highways which lead from the channel ports of Kent, welcoming every traveller home.

(plate 47)

Winter Heliotrope
Petasites fragrans (Vill.) C. **Presl.**

Coltsfoot
Tussilago farfara L.

Mugwort
Artemisia vulgaris L.

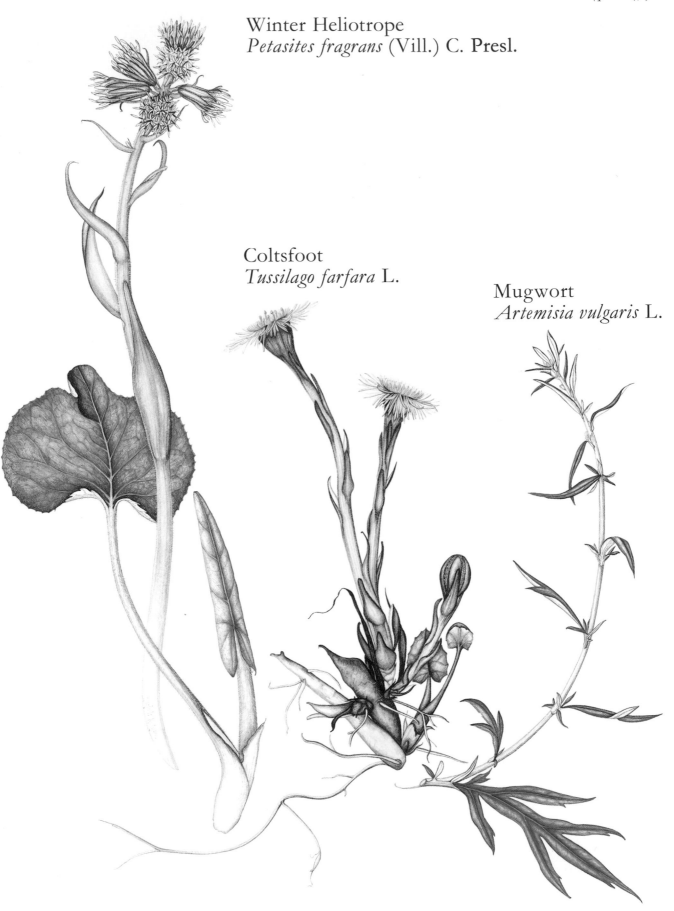

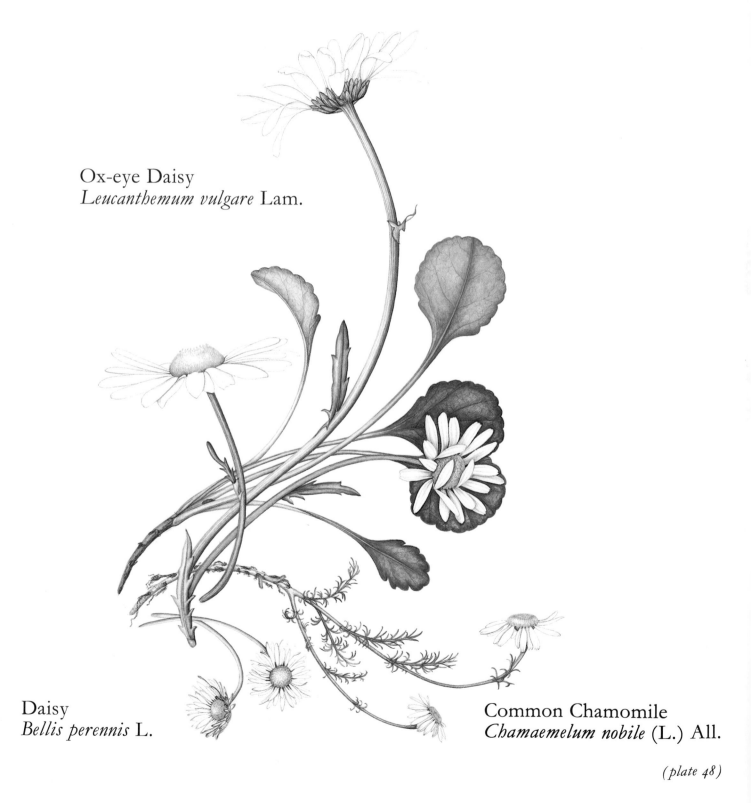

Ox-eye Daisy
Leucanthemum vulgare Lam.

Daisy
Bellis perennis L.

Common Chamomile
Chamaemelum nobile (L.) All.

(plate 48)

Winter Aconite *(plate 35)*
Eranthis hyemalis (L.) Salisb.

The earliest flowering member of its family, it was imported into British gardens from southern Europe where it graces the woodland edges and mountain meadows in spring. The flowers, which are surrounded by three frilly leaves, looking like the ruff around a choirboy's upturned face, are very sensitive to temperature, opening as soon as the temperature rises above 10°C (50°F). A very useful addition to any spring garden, but it can become a weedy nuisance.

Apart from its angelic appearance it has no recorded virtues, for it has a burning taste and is very poisonous.

Meadow Buttercup *(plate 35)*
Ranunculus acris L.

A buttercup without runners and an unchannelled flower-stalk. Like the last, it has an acrid burning taste; and like all the other buttercups is poisonous probably due to the presence of the alkaloid anemonin. Some people are allergic to the plant and will blister if they handle it in quantity. Gerard makes the same point with a little more acrimony:

Many do use to tie a little of the herbe stamped with salt unto any of the fingers, against the pain of the teeth—which medicine seldome faileth; for it causeth greater paine in the finger that was in the tooth, by the meanes whereof the greater paine taketh away the lesser . . .

Cunning beggars do use to stampe the leaves, and lay it upon their legs and arms, which causeth such filthy ulcers as we dayly see (among such wicked vagabonds) to move the people the more to pitie.

Creeping Buttercup
Ranunculus repens L.

A buttercup with runners and a furrowed flower stalk, it is thus able to run all over the fields rooting as it goes, and furrow the farmer's brow, for it is a noxious weed avoided by cattle. However like all buttercups it loses its acridity when dry and so does not spoil the hay, and may well help us all to enjoy good butter throughout the winter.

It is strange how so many people only look at the flower, taking no note of the way in which the arching runners root at each node which also produces a new tuft of leaves. It is indeed that tufted habit that made the classical scholars believe that this is the tufted-crow-toe of which Milton writes.

Bulbous Buttercup
Ranunculus bulbosus L.

A buttercup arising from a small bulbous swelling or corm-like tuber at the base of the plant, with furrowed hairy flower stalks, and sepals reflexed (turned back) as if in aversion to the gold of the petals. Of the three common buttercups, this is the earliest to bloom and it chooses the driest places in which to grow. In fields which were ridged like corrugated iron by early attempts at drainage, the ridges turn gold in May due to the abundance of this plant, the furrows only golding later with their own abundance of the other species. Though also called St Anthony's Turnip, it is both emetic and poisonous unless dried or boiled. Despite this, pigs feed on it avidly even when fresh.

CLASS 14: ORDER 1
MINT FAMILY (see also page 135)
LABIATAE

An important family which puts flavour into our lives as it has for centuries in the form of mint, thyme, balm, basil, savory, clary, hyssop and marjoram.

Mints
Mentha spp.

What would spring lamb be like without mint sauce, dinner without its After Eights? Despite the fact that the Lord High Executioner got dyspeptic when it was puffed in his face, today the flavour is ruminated upon by millions.

Corn Mint
Mentha arvensis L.

Water Mint
Mentha aquatica L.

Bushy Garden Mint
Mentha × *gentilis* L.

Mitcham in Surrey supplied all London's mint for more than a hundred years, and Mitcham Mints were the strongest you could buy, so said my granny. Despite its gastronomic and medicinal virtues there is another which does appear to work. The leaves of Water Mint laid in the larder not only impart their own special smell to that place, but also ward off mice.

Self-heal
Prunella vulgaris L.

A common plant with astringent properties and used to stem bleeding caused in the harvest field, hence its other names Sicklewort and Hook Weed.

Hedge Woundwort *(plate 36)*
Stachys sylvatica L.

Another useful plant in cases of cuts; for placed on the wound the hairiness of the leaf helps in the coagulation of the blood. Don't be put off by the malodour, it really works.

Black Horehound *(plate 2)*
Ballota nigra L. ssp. *foetida* Hayek

Subspecies *foetida* means that this smells real bad. Despite and perhaps because of this, it was used both as an infusion with honey and as horehound candy for coughs and colds; and some say it is worthy of much more respect and research even in these days of synthetics.

This plant, which enjoys the company of man, has made use of our colonising ability along with its own, to spread across the world.

Field Pansy
Viola arvensis Murray

Common Violet
Viola riviniana Rchb.

(*plate 49*)

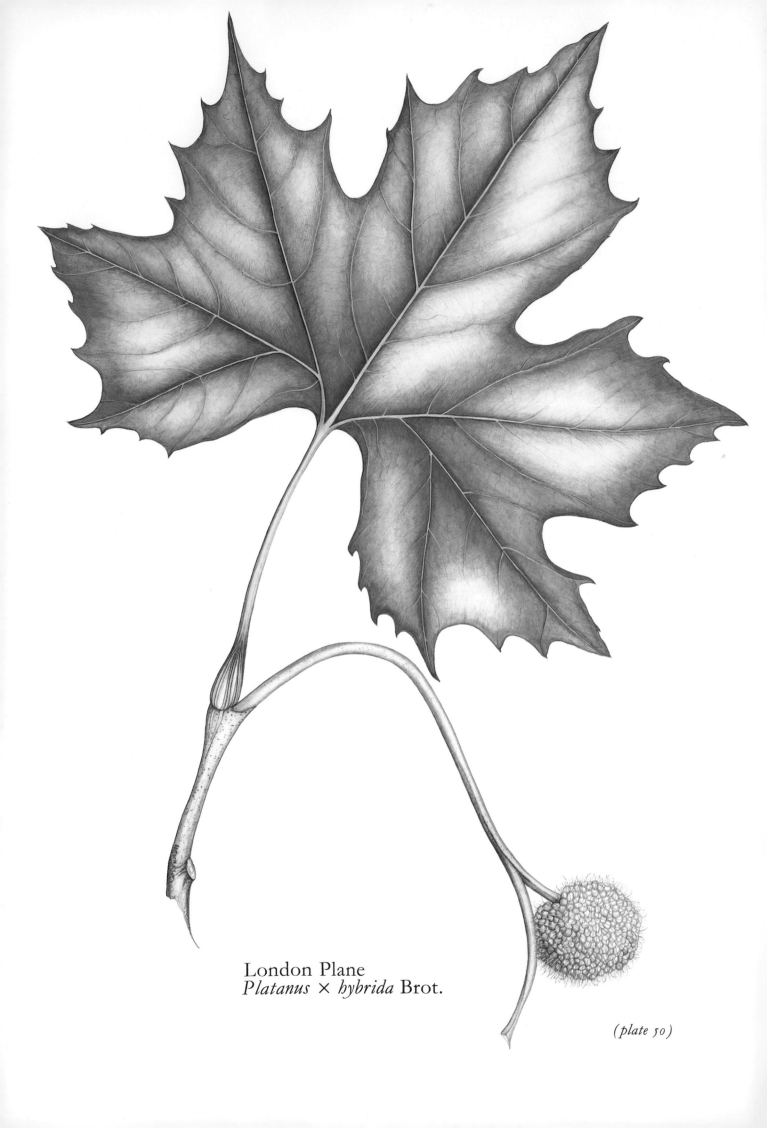

London Plane
Platanus × *hybrida* Brot.

(plate 50)

Dead-nettles
Lamium spp.

Although they look rather like their notorious counterparts in a separate family, these do not sting.

Henbit
Lamium amplexicaule L.

Red Dead-nettle *(plate 36)*
Lamium purpureum L.

White Dead-nettle *(plate 36)*
Lamium album L.

The leaves of all may be collected in spring and used as a pot herb.

Hemp-nettle
Galeopsis bifida Boenn.

This plant was perhaps produced by a mixing of the genes brought about when a hemp-nettle from the continent was introduced and grown near the English large-flowered hemp-nettle. We know that plants resembling the common hemp-nettle can be produced in this way, and a very handsome plant it can be.

Ground Ivy
Glechoma hederacea L.

This was sold in Queen Elizabeth I's time as Gill-by-the-ground, Hay Maid, Ale Hoof and many other names. It was mainly used in pulmonary disorders and in disorders of the eye both in man and horses. 'Hoof' from the shape of its leaf, and 'Ale' from its curing of all ills.

Bugle *(plate 2)*
Ajuga reptans L.

Used both for cuts and for coughs, its deep-blue spikes of flowers are a worthy herald of summer.

CLASS 14: ORDER 2
FIGWORT FAMILY (see also pages 133, 138 and 141)

Common (Knotted) Figwort *(plate 37)*
Scrophularia nodosa L.

Nodosa and 'knotted' come from the swellings (knots) on the roots which vary from the size of a pea to that of a broad bean, and were looked upon as signatures to its main virtue—the cure of scrofula, the 'king's evil'. And all you had to do was hang it about your neck.

Foxglove *(plate 37)*
Digitalis purpurea L.

A very beautiful plant which is deadly poisonous, for all its parts contain digitalin, an explosive heart stimulant. It was, and still is, much used in medicine; but extreme care has to be taken with the dosage.

The flowers have always delighted children, and to trap a visiting bee has long provided excitement on a country walk, the bravest of the party having to decide just when to let go to the mouth of the flower and release its pollen-covered captive. The leaves were used to darken the lines etched into the stone floors of houseproud people.

Poisonous as this plant is to us, careful inspection of its leaves will show that a number of insects and slugs are immune, and can take a hearty meal from its leaves.

CLASS 15
CABBAGE FAMILY   CRUCIFERAE

CLASS 15: ORDER 1
Wart-cress (Swine-cress)
Coronopus squamatus (Forskal) Ascherson

Slender Wart-cress (Lesser Swine-cress)
Coronopus didymus (L.) SM.

The first is native, the second introduced; but both are common countryside weeds, the former thriving best in trampled ground around gates and feeding-troughs. The introduction came probably from South America, perhaps brought in with potatoes, tomatoes and the other more useful garden plants which also originated there. They differ not only in size but by the notching of their pouch-like fruits.

Shepherd's Purse *(plate 38)*
Capsella bursa-pastoris (L.) Medicus

The first two names mean the same thing and refer to the notched pouch-like fruit which holds 12 seeds in each compartment. It is one of the most variable plants, an annual or biennial herb which may flower and fruit be it 3cm (1in) or 40cm (15in) tall. A truly cosmopolitan weed found in every continent and so common that it was used in all forms of medicine. It was at one time known as St James's Wort or Poor Man's Pharmacetie. Washed, lightly boiled, preferably without its capsules, it makes a tasty green vegetable.

CLASS 15: ORDER 2
Cabbage
Brassica oleracea L.

The mother's lament of 'Be a good boy or girl and eat up your greens', links two facts—the importance of green vegetables to our diet, and the British fad of overcooking and hence ruining this king of vegetables both gastronomically and dietetically.

In its wild form a straggling plant with an elongate flowerhead, probably native along our coasts which have always offered open habitats for would-be weeds. Plant fanciers and breeders across the ages have reached into the genetic diversity of its wild populations across Europe and produced delicious things like cabbages of all sizes, white and red, Brussels sprouts, savoys and broccoli.

Chinese Mustard
Brassica juncea (L.) Czern.

A native of Asia, it is cultivated for the oil in its seeds of which 8–12 are produced in each cell of an elongate fruit

(plate 51)

Reedmace
Typha latifolia L.

which may be 5cm (2in) long. This is one of the plants which helps to turn so much of our best cropland yellow each summer, and draw the honey bees away from orchards and their more natural source of flavourful nectar.

Charlock *(plate 38)*
Sinapis arvensis L.

A once common, perhaps the most common, weed of cornfields and an early victim of the selective herbicides which have done so much to help farmers wage war against the weeds. Charlock goes under many common names, from Bread and Marmalade to Runch-balls and Wild Mustard.

White Mustard
Sinapis alba L.

What would our meat be like without the gold of mustard, be it French or English? In the preparation of the former, the dark husks are not separated from the seeds, hence the less bright colour and the stronger taste—so strong that it is often diluted with vinegar and must (new wine), hence *mustun ardens* ('hot must'). The seeds are also a must with children, providing the most instant of gardening; and the product sandwiched between bread provides a problem in garden-party etiquette.

Wild Radish
Raphanus raphanistrum L.

The swollen tap root of the cultivated radish would never be served in polite garden-party circles for fear of repetition. Its wild cousin reminds us of the fact, for its long jointed fruit looks as if it has been inflated at intervals by some ill wind. The botanical name for such a fruit is a lomentum, as if lamenting the fact. The sweet acrid taste of the cultivated cousin, whose origins are not yet understood, serves to remind us again and again of the unique chemistry of the plants of this Cabbage Family, which has given our taste buds so much to enjoy or to complain about.

Wavy Bitter-cress
Cardamine flexuosa With.

Hairy Bitter-cress
Cardamine hirsuta L.

These two Cresses are found growing about the cascade at the end of the lake in the Palace Gardens—a typical habitat for the former, less so for the latter which can thrive in dry situations. The long fruits, especially of the latter, are explosive—coiling violently when touched, flinging out their seeds.

Creeping Yellow-cress
Rorippa sylvestris (L.) Besser

Marsh Yellow-cress
Rorippa islandica (Oeder) Borbás.

The genus *Rorippa* includes Watercress, which finds its way into sandwiches along with mustard, which grows up under the name of cress seeds. Very confusing, that's why it is best to stick to Latin names, for each plant has only one, officially.

Cultivated since 1808 around London, the Summer Watercress should be called *Rorippa nasturtium-aquaticum* (L.) Hayek. The Winter Watercress is a hybrid, *Rorippa × sterilis*. The × denotes a hybrid or bastard, and *sterilis* the fact that only about 1 per cent of the seed it produces is viable. No matter to the watercress gardener, for both forms are propagated by cuttings.

Hedge Mustard
Sisymbrium officinale (L.) Scop.

The fruit (siliqua) of this plant is very long in relation to its width, the long tail of the letter Q helping one to distinguish it from the short fat fruits (siliculae) of some other members of this Cabbage Family.

Not the most famous member of the genus to be found in London; that distinction goes to *Sisymrium irio* L., London Rocket, for this plant evidently covered the city after the Great Fire in 1666. Since that time many observations have indicated that this plant thrives on the presence of ash added to its habitat. The *officinale* of Hedge Mustard denotes that it is of medicinal use as a cure for hoarseness of the throat, and it was known in France as *Herbe aux chanteurs*.

Thale Cress
Arabidopsis thaliana (L.) Heynh.

If you want perfection in miniature this is a plant to look out for growing on dry walls where it may flower even when no more than 2cm ($\frac{3}{4}$in) tall. Such tiny flowers produce their tiny siliquae, each valve of which contains at least 25 seeds. Take these and grow them in a well-prepared plot and the progeny will probably reach at least 50cm (20in) tall, again with tiny flowers and tiny pods each containing upwards of 70 seeds.

CLASS 16: ORDER 5

GERANIUM FAMILY   GERANIACEAE
Dove's-foot Cranesbill
Geranium molle L.

Everyone knows a geranium, thanks to the garden plants which go by that name. Unfortunately these ever so common pot-plants have asymmetrical flowers and hence are not geraniums but pelargoniums. The perfectly 'clock-faced' flowers of our wild geraniums are pure symmetry, and if one were sliced in two at any angle it would provide two equal halves. The name Cranesbill comes from the shape of the fruit which is like a long beak, and *geranos* is Greek for crane.

To further confuse the avian issue this particular plant is called Dove's-foot, from the form of its leaves which are soft and downy like velvet and of a greyish blue-green colour much more like a Pigeon, and indeed in France it goes under the name Pied de Pigeon. Some would of course say that Doves and Pigeons are one and the same.

Like its somewhat commoner cousin Herb Robert, it flowers from March through August and into September. The latter however has a very obnoxious smell, as do most of the species if you handle their leaves. This scent is said to ward off insects, and by Culpeper to cool a hot brain.

CLASS 17: ORDER 2
FUMITORY FAMILY   FUMARIACEAE

Common Fumitory
Fumaria officinalis L.

Not the first plant in these Royal Lists to bear a title translatable as 'official', meaning that it is of medicinal use. And if we turn to Culpeper's *Pharmacopoeia Londinensis*, 1675 edition, we find:

Fumitory. Cold and dry, it openeth and cleaneth the urin, helps such as are Itchey and Scabby, clears the Skin, opens stoppings of the Liver and Spleen, helps Rickets, Hypochondriak Melancholy, Madness, Frenzies, quartan Agues, Loosneth of the Belly, gently purgeth Melancholy, and adult Cholera: boyl in white Wine and take this one general rule, All things of a cleansing and opening nature may be most commodiously boiled in white wine. *Remember but this, and then I need not write one thing so often.*

A veritable national health service right in your own back yard and much more, for the name *Fumaria* is probably derived from the fact that its smoke was used in exorcism.

CLASS 17: ORDER 4
PEA FLOWER FAMILY   FABACEAE

The family which gave us all the diversity of the peas and beans including both those now fabled plants baked and soya. The family which has also given some hope to the third (or rather two-thirds) of the world who live on the brink of protein malnutrition, for the seeds are rich in amino acids.

Laburnum
Laburnum anagyroides Medicus

Like many of the members of this large and important family its fruit is a legume and unfortunately looks like a bean or pea pod—unfortunately, because its pea-like seeds are poisonous. Its great sprays of oh-so-yellow flowers are ample reward for the timely warning which should be repeated every year. Another name for it is Golden Rain, and another feature which makes the fruits of interest to the young is their explosive dehiscence. Introduced from central Europe, no part of it should even be tasted for all contain cystine and laburnine both of which are deadly.

Broom *(plate 39)*
Sarothamnus scoparius (L.) Wimmer ex Koch

Broom got its name from its use in the manufacture of besoms, and its connection with the black arts perhaps through their supposed use as the personal transport of witches. So we find this cautionary verse:

If you sweep the house with blossomed broom in May
You are sure to sweep the head of the house away.

It is also said that Count Geoffrey of Anjou, admiring the tenacious quality of broom, picked a sprig and rode into battle crying 'Planta genista, Planta genista', way back in 1140. After the victory he adopted its name for his family who carried it until the death of Richard II. The Plantagenet crest was a genet passing between two sprigs of broom.

Red Clover *(plate 40)*
Trifolium pratense L.

Alsike *(plate 40)*
Trifolium hybridum L.

White Clover *(plate 40)*
Trifolium repens L.

Clover gets its common name from the Latin word for 'club', the three leaflets being like the triple-headed club of Hercules. St Patrick used such a leaf not as a vegetable but as a living parable to instruct the heathen Irish in the doctrine of the Trinity. Whether the original Shamrock growing on the shores of County Wicklow where he landed was a Clover or Wood Sorrel is still debated, and the fact that it was eaten by the peasants of that country during the hard days of famine does not help the matter, for the leaves of both are edible. Both red and white clover are native, but the alsike was introduced from Europe and has been cultivated since the eighteenth century.

The importance of clover as a crop for improving the nutrient status of the soil was realised long before the discovery that the root nodules are homes for symbiotic bacteria which fix nitrogen from the atmosphere and add it to the soil in the form of nitrate. This important property, which is found in many members of the family, has helped to make such nodulated plants effective weeds, and hence a real menace, in many countries. No wonder though that farmers like clover and so do bees.

> The pedigree of Honey
> Does not concern the Bee
> A Clover any time to him
> Is aristocracy.
>
> Emily Dickinson

Bird's-foot Trefoil *(plate 41)*
Lotus corniculatus L.

A common plant with many common names—Bacon and Eggs on account of the colour of its buds and flowers, Shoes and Stockings from the shape of the same; Butter Jogs and Cross Toes are perhaps more obscure. The latter is probably from the way the fruits, which form a bird's foot, cross over when drying, or perhaps that those same pods have divisions between the seeds. The massed presence of this plant fills the air with a sweet though not sickly fragrance which signifies summer is in full swing.

Stinging Nettle
Urtica dioica L.

Small Nettle
Urtica urens L.

(plate 52)

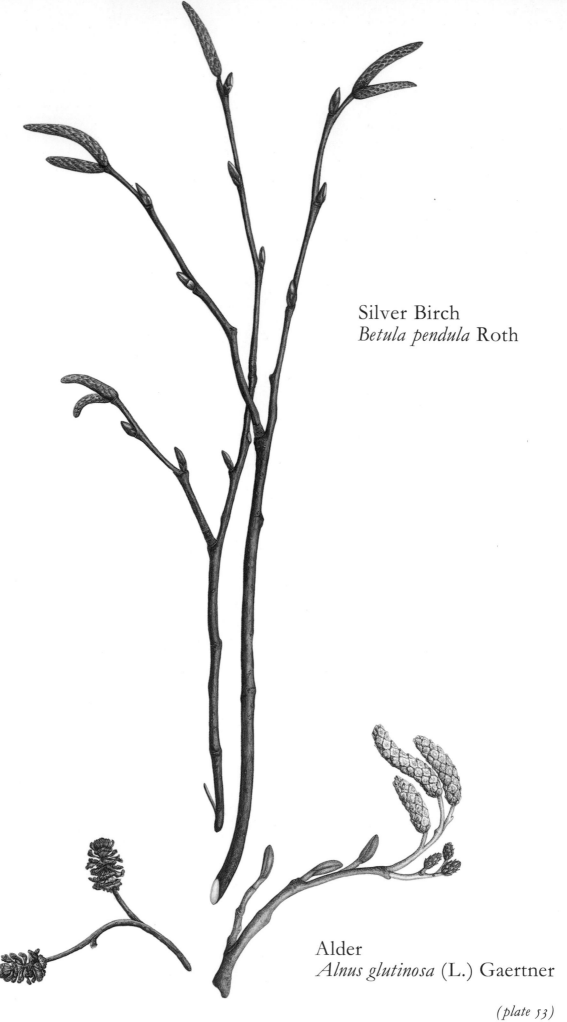

Silver Birch
Betula pendula Roth

Alder
Alnus glutinosa (L.) Gaertner

(plate 53)

False Acacia
Robinia pseudacacia L.

The oh-so-suburban Acacia Avenue gets its name from this tree which was introduced to this country from North America. The tree itself gets its names from a Monsieur Robin who first introduced it to Europe and the fact that it is not an Acacia. It was William Cobbett who popularised it under its other common name of Locust. Perhaps the true Locust Tree comes from the Mediterranean where its large bean-like seeds, carobs, were used to determine the weight of diamonds—hence carats.

False acacia is a good lawn tree from its bark which is like gothic tracery through its delicate compound leaves to its hanging flower masses and persistent deep-red pods. In the garden unless tended it tends to sucker, producing, if allowed, a thicket of very spiny shrubs.

Common Bird's-foot
Ornithopus perpusillus L.

The white-cream flowers each veined with red are borne in clusters of 3 to 6, and as they turn into strongly jointed curved pods the whole looks very much like the foot of a bird. Being the smallest of our plants in this family, a lens will be required in order to appreciate its full beauty. The plant does its best to help you, for each flowerhead is presented on a pinnate leafy bract.

Smooth Tare
Vicia tetrasperma (L.) Schreber

Tufted Vetch *(plate 39)*
Vicia cracca L.

The Vetches or Tares have compound leaves made up of many leaflets, but never have a leaflet at the top, though they usually have tendrils which may be branched.

The smooth tare has no hairs and but 4 seeds in each pod. The tufted vetch, which may grow as tall as you or I, bears great masses of blue-purple flowers as it uses its tendrils to clamber up any support. It is especially at home in the hedgerows where it provides welcome and nutritious grazing for stock and an abundance of seeds for birds. However, like the tares of the Bible, it can fall on the best-prepared ground with less pleasing results. Since the days when Plot sang the agriculture praises of the vetches in his *History of Staffordshire* (1666), many have suggested that fields be put down to these fast-growing, nutritive, soil-conditioning plants. The only problem is that in their growth they would strangle each other; one reason why the Medicagos with their more upright less clinging habits, for they lack tendrils, have provided the farmer with the lucernes or alfalfas.

Meadow Vetchling (Meadow Pea) *(plate 41)*
Lathyrus pratensis L.

Greater Everlasting Pea *(plate 41)*
Lathyrus latifolius L.

Winged or angled stems help to separate the peas from the vetches—beautiful plants without which our hedgerows,

roadsides and waste patches would lose much of their summer beauty. In the last century, botanists sang the praises of such plants, but warned would-be 'improvers' of the countryside against broadcasting seed of alien plants into areas of natural or semi-natural vegetation. The same is even more true today, for now there is much less of our wild countryside left. However, the planting of such alien seeds in wild corners of our gardens can bring not only us, but the local birds and bees, much benefit.

CLASS 18: ORDER 4
ST JOHN'S WORT FAMILY HYPERICACEAE

Wort is an Old English word for common plant or herb, and we have ample evidence of its early presence in London for Stow writes in his *Survey of London* (1598): 'On the Vigil of St John [that is Midsummer Eve] every man's door was shadowed by green birch, fennel, St John's wort, orpine, white lilies and the like.' Whether it was efficacious in warding off evil spirits we can only guess, but a balm or salve made from its flowers was used as a treatment for cuts and scratches well into this century in the more rural parts of Kent.

Common St John's Wort
Hypericum perforatum L.

Square-stalked St John's Wort
Hypericum tetrapterum Fres

Trailing St John's Wort
Hypericum humifusum L.

The leaves of all three are furnished with small translucent oil-producing glands. They are however most abundant—as its name signifies—in the first; for on holding it up to the light the leaves look as if they are perforated. Perhaps this was the link with St John the Baptist—the likeness of the perforations to wounds. The flowers of all leave a yellow stain upon our fingers, and may be mordanted with alum to produce a stable yellow dye.

CLASS 19
DAISY FAMILY COMPOSITAE

'Compositae' from the composite flower which consists of many flowers arranged with geometrical precision, as described by the thirteenth-century mathematician Fibbonacci, on a capitulum or head.

This largest and most successful family of the flowering plants has provided us with many very useful things—lettuce, endive, Jerusalem artichoke, globe artichoke, salsify, chicory, scorzonera and sunflower seeds and oil—and just as many weeds.

CLASS 19: ORDER 1: GROUP 1
Nipplewort *(plate 42)*
Lapsana communis L.

Hornbeam
Carpinus betulus L.

Hazel
Corylus avellana L.

(plate 54)

Beech
Fagus sylvatica L.

(plate 55)

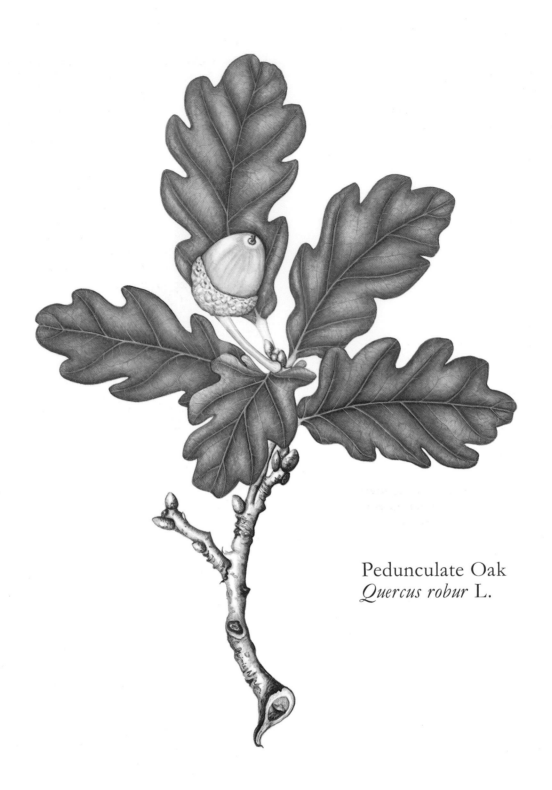

Pedunculate Oak
Quercus robur L.

(plate 56)

Formerly used as a salad plant, the shape of its closed flowers gives it its common name, and its supposed medicinal properties.

Common Cat's Ear
Hypochaeris radicata L.

The radical leaves with prominent hairs lying flat against the ground provide it with its name, and protection from the scythe and the attention of most grazing animals. However, pigs enjoy it, and will root it up with great effect.

Autumn Hawkbit
Leontodon autumnalis L.

Too common to say much about, except that its hairs are each forked—take a look through a lens.

Lesser Goat's-beard *(plate 42)*
Tragopogon pratensis L.
ssp. *minor* (Mill.) Wahlenb.

The flower and fruiting heads close each day around noon, hence its other name Jack-go-to-bed-at-noon. A close relative of Salsify—a vegetable well worth growing—its roots may be eaten boiled in water until tender, and are preferred by some.

Prickly Lettuce
Lactuca serriola L.

Probably native, this wild lettuce has narcotic properties and, if a large amount is eaten before rest, induces sleep. The origins of the cultivated lettuces of today are somewhat obscure, but the narcotic element has been to a large extent bred out.

We know that Henry VIII liked lettuce, and that in 1530 the gardener of York Place received a reward 'for bringing lettuze and cherries to Hampton Court'. It is still highly prized, for to produce lettuce under glass in Britain during the winter of 1971–2 required 500 units of energy (fossil fuel) for every unit of energy (edible lettuce) produced—a fantastic investment? I bet it gave the farmer a few sleepless nights, or perhaps he ate some of the produce. At the price he had to charge to make a profit, I bet he didn't eat enough to make him sleep.

Corn Sow-thistle *(plate 43)*
Sonchus arvensis L.

Smooth Sow-thistle *(plate 43)*
Sonchus oleraceus L.

Prickly Sow-thistle
Sonchus asper (L.) Hill

Three native plants now common weeds across the world. The leaves of all three boiled and prepared in the way of spinach were said to furnish a dish of vegetables superior to any green plant in use. And that statement was made in France.

The Queen's Own Hawkweed *(plate 44)*
Hieracium lepidulum (Stenström) Omang
var. *haematophyllum* Dahlst.

I make no apologies for the Royal common name. The fact is that the hawkweeds are apomictic, which means they can produce seeds without fertilization. In essence this is equivalent to virgin birth, and hence the offspring are identical to their parents. So once a new line is established by mutation or chance, or cross-fertilization, it becomes fixed—a new 'species'. No less than 260 such species have been recognised in Britain, but their identification is best left to the expert and so most of them do not have common names.

This variety of a species which was introduced into Britain from Europe has more densely glandular heads and red-tinted leaves, and so may be a new 'apomictic species' in the making. We have this firm record from Buckingham Palace where it is now possibly extinct, and must watch for other records across Britain. I feel the name The Queen's Own Hawkweed is therefore appropriate. The 'species' and this variety are spreading along railway and roadside banks, so watch out—this Royal visitor may soon be up or down your way; but only the experts will really know; check yours against Plate 44.

Beaked Hawk's-beard
Crepis vesicaria L.
ssp. *taraxacifolia* (Thuill.) Thell.

Smooth Hawk's-beard
Crepis capillaris (L.) Wallr.

The first was introduced from the Continent, the second is a native. Both are annuals but can be biennial or even perennial, and are thus well fitted to the weedy way of life.

Dandelion *(plate 45)*
Taraxacum officinale Weber

Its leaves may be eaten raw or after blanching, its root makes a good cup of coffee which instils sleep, not wakefulness, and its clock is used to pass the time of day. Its diuretic properties which give it other names like Pee-a-bed and Wet-a-bed have medicinal truth, although it must be taken internally, not just handled.

The white latex in the stalk is a form of rubber, and during World War II Russia cultivated many thousands of acres of a species especially rich in latex. A gold-white harvest of this weed provided the war machine of that country with much of the rubber it required to roll smoothly forward to victory along with her then allies.

CLASS 19: ORDER 1: GROUP 2

Lesser Burdock *(plate 46)*
Arctium minus Bernh.

Aptly named, this much maligned weed which is distributed on the socks of boys and girls can provide a very wholesome vegetable. The young stems should be stripped of their rind just before flowering, the succulent inner part being prepared like asparagus.

Scots Pine
Pinus sylvestris L.

(plate 57)

Wayside Cudweed
Gnaphalium uliginosum L.

One of the commonest of our everlasting flowers, for its white chaffy bracts do just that. Gnaphalium means 'soft down', and this covers its leaves.

Mugwort *(plate 47)*
Artemisia vulgaris L.

A plant which has long been used in herbal baths; and travellers were wont to lay it in their shoes where, if it did not give them strength, it certainly cleared the air when they removed their footwear at the end of a hard day.

If you are removing mugwort, look for the small black dried roots which were of old called 'coals' and which were reputedly endowed with many medicinal properties.

CLASS 19: ORDER 2: GROUP 2

Gallant Soldier *(plate 43)*
Galinsoga parviflora Cav.

Shaggy Gallant Soldier
Galinsoga ciliata (Rafin) Blake

Both introduced—the former from Peru, the latter from Mexico or South America—and both now very much at home.

The first-named is also known as Joey Hooker after Sir William's son, who perhaps first recorded it at Kew.

Niger
Guizotia abyssinica (L.f.) Cass.

Introduced from Africa where the oil from its seeds is used—as it is in India and the West Indies—for food, burning oil and in the manufacture of soap.

THE RAGWORTS   SENECIO

Common Ragwort
Senecio jacobaea L.

This plant is poisonous to cattle if eaten in quantity. The animals avoid eating it if they can, thus helping it to establish itself once it has a roothold.

Oxford Ragwort *(plate 43)*
Senecio squalidus L.

This plant, which feeds the Cinnabar Moth caterpillars, is the commonest flower of London streets and provides instant beauty on roadworks and the like. It was introduced from Italy and its exact origin is probably from the flanks of Vesuvius, so no wonder it does so well on town pavements. It was first recorded on Oxford walls in 1794.

Sticky (Stinking) Groundsel *(plate 42)*
Senecio viscosus L.

A doubtfully native plant which lives up to both its common names. It loves disused railway embankments and has spread since Dr Beeching's cutbacks.

Common Groundsel
Senecio vulgaris L.

A native, the leaves of which boiled with wine or water were said to cure stomach ache. Please, if you live near farms which grow Sugar or Red Beet, get rid of it for the farmers' sake, for it harbours a disease of their crop.

Golden Woundwort
Senecio doria L.

Introduced from Europe, but established in only a few places, the Palace Gardens being one of them. It is a pretty plant and reminds us how this genus has given us plants of great beauty like the Cinerarias.

Coltsfoot *(plate 47)*
Tussilago farfara L.

The flowers come up first, attracting flies and bees; they close at night and make good wine. The leaves come only later and are known as Poor Man's Baccy, for they can be smoked. They were also said to help asthmatics.

Winter Heliotrope *(plate 47)*
Petasites fragrans (Vill.) C. Presl.

Introduced from the western Mediterranean, the sweet-scented flowers which appear in January attract flies and hive bees. Its underground stem is creeping, and a devil to eradicate.

Common Fleabane
Pulicaria dysenterica (L.) Bernh.

This enjoys wet places, was burned to expel fleas and other wildlife from homes and barracks, and was used to treat dysentery. If we ever want to send a plant into space this is a good candidate, for it has been shown that it is able to live in a complete vacuum or in pure nitrogen for as long as six months.

Garden Golden-rod
Solidago altissima L.

Early Golden-rod
Solidago gigantea Aiton

Although we do have a native golden-rod, those which have been much planted in gardens and have escaped to become naturalised mainly originated in North America.

At one time it was so highly esteemed in the stemming of blood from wounds that it was imported dried from the Continent. Gerard writes:

It is esteemed above all the herbs for the stopping of blood ... within my remembrance I have known the drie herb, which came from beyond the seas sold in Bucklersburie in London for half-a-crown an ounce. But since it was found in Hampstead Wood, even as it were at our towne's end, no man will give half-a-crown for an hundred weight of it; which plainly setteth forth our inconsistencie and sudden mutabilitie, esteeming no longer of anything, how precious so ever it be, the whilst it is not strange and rare.

How true this still is today!

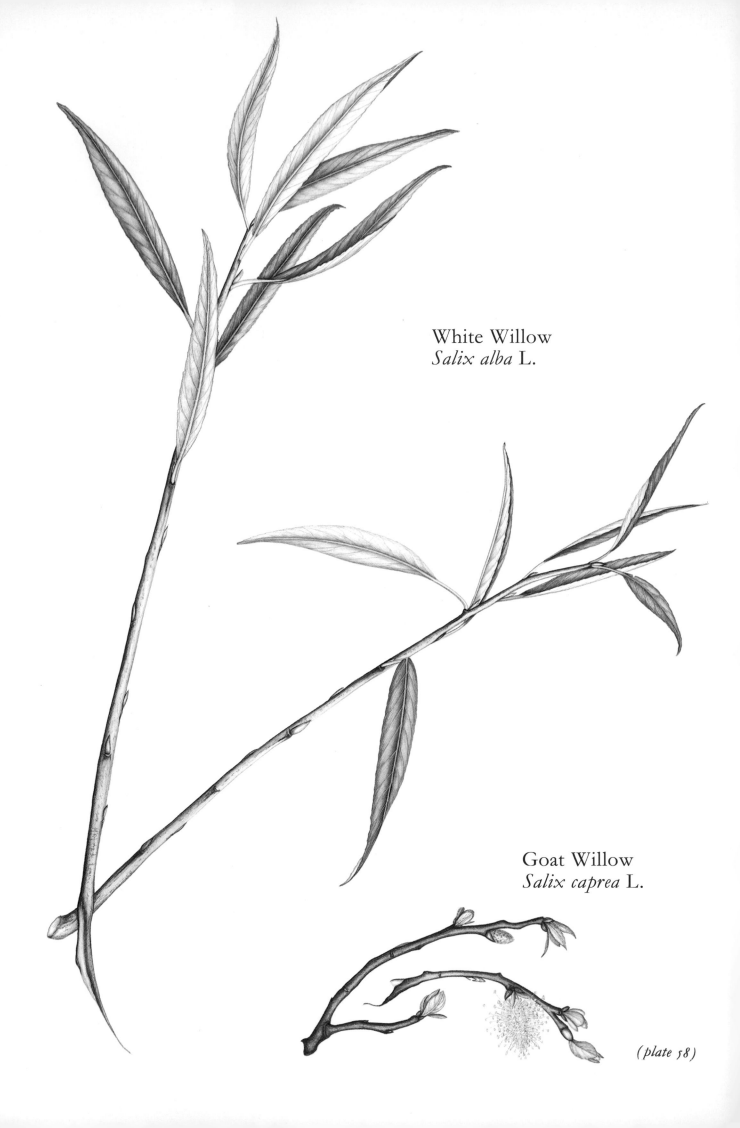

White Willow
Salix alba L.

Goat Willow
Salix caprea L.

(plate 58)

(plate 59)

Yew
Taxus baccata L.

Holly
Ilex aquifolium L.

(plate 60)

Michaelmas Daisy
Aster novi-belgii L.

First brought from America by the Tradescants and grown in their garden in Lambeth, the many forms have moved out and become naturalised in many places. It was to A. *lanceolatus* Willd. that the name Michaelmas daisy was first applied.

Canadian Fleabane
Conyza canadensis (L.) Cronq.

Introduced from Canada via Paris, it can become a bane to gardeners but not to fleas.

Daisy *(plate 48)*
Bellis perennis L.

What more need be said except 'she loves me, she loves me not'—a sentiment that rings true in any garden with a well-kept lawn.

Common Chamomile *(plate 48)*
Chamaemelum nobile (L.) All.

When you tread on the Palace lawn, you walk on history and upon this plant.

Stinking Chamomile
Anthemis cotula L.

Its truly horrible smell and the fact that it blisters the hands of reapers, has led to its being given several disgusting names. It was used at one time as an insect repellant, as were many others of this family; and, as already reported, a close relative which lives in Africa and Asia produces the really safe insecticide pyrethrum.

Yarrow
Achillea millefolium L.

Achilles, no less, was the first to use this as a wound wort. It was also at one time used as a condiment, for another name is Old Man's Pepper, and in the Orkneys as a tea for dispelling melancholy. It is a very common plant and very variable, so that you may come across it in any size from a small though not stunted specimen in full flower on our mountain tops to great masses more than 60cm (24in) tall in our lowland fields.

Scentless Mayweed
Tripleurospermum maritimum L.
ssp. *inodorum* (L.) Hyl. ex Vaarama

Three strong ribs on the posterior face of the fruits give this its long generic name. Each plant may produce between 10,000 and 210,000 seeds. Put that in your garden computer and buy a hoe.

Ox-eye Daisy *(plate 48)*
Leucanthemum vulgare Lam.

Also called the Dog Daisy, the reason being more than obvious once you have walked through a field of these flowers. Two Dog Daisies came from America, the other from Japan; and they were used together with a Mi-chaelmas Daisy from Britain to produce the truly fabulous dazzlingly white Mount Shasta Daisy. The plant breeder was Luther Burbank and it took more than eight years and selection from millions of offspring to achieve this Queen of Daisies. What other wonders wait to be discovered in the gene banks of the world's 'weeds'—gene banks which we are now wantonly destroying?

Pineapple Weed
Matricaria matricarioides (Less.) Porter

Introduced probably from north-east Asia, it lives up to its name, for the whole plant including its yellow mound-like rayless flowers smells strongly of pineapple when crushed.

Scented Mayweed
Matricaria recutita L.

Used as a substitute for real Chamomile, the oil being distilled from its flowerheads.

Spear Thistle *(plate 46)*
Cirsium vulgare (Savi) Ten.

Creeping Thistle
Cirsium arvense (L.) Scop.

Really noxious weeds, and I can say this because neither of these is the emblem of Scotland. The insignia borne by the Knights of the Order of the Thistle is a gold collar with sprigs of another thistle and rue interlaced.

Before the introduction of turnips, which revolutionised farming in Scotland, cattle were suppered with the unlikely repast of thistles for five or six weeks each year. In lighter vein, a cushion filled with thistledown is fit for a King or Queen.

CLASS 19: ORDER 3
Hardheads
Centaurea nigra L.

Slender Hardhead *(plate 46)*
Centaurea nigra ssp. *nemoralis* (Jord.) Gugl.

The red purple of its bracts and the hard heads which remain are so much a part of summer that its weedy properties should be forgiven if not forgotten. It is however a pernicious weed of grazing land and remains untouched by cattle either when fresh or dried.

Mediterranean Knapweed
Centaurea diluta Aiton

Imported from the Mediterranean but definitely 'baggage not wanted on voyage'.

CLASS 19: ORDER 6
VIOLET FAMILY   VIOLACEAE
Common Violet *(plate 49)*
Viola riviniana Rchb.

One of our common violets, but without a sweet scent. It can be distinguished from the Pale Wood Violet with

(plate 61)

Common Horsetail
Equisetum arvense L.

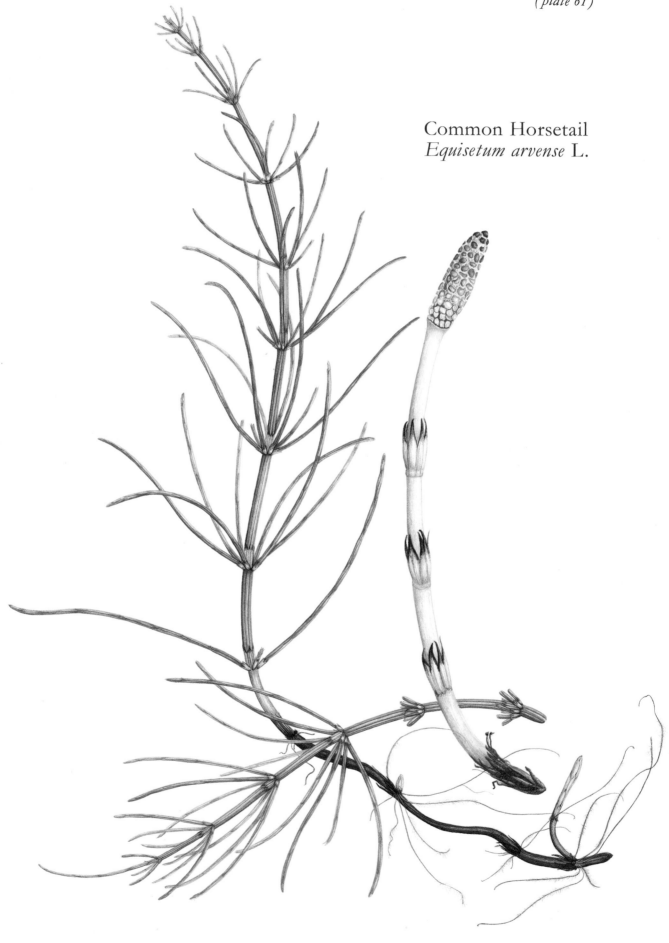

which it often grows by the fact that the spur of the flower has a notched apex rather like the dent in the top of a trilby hat. The capsule is also of great interest and well worth watching. As it ripens it is lifted upright on its stalk, gaining height in so doing, then splitting into three valves it opens to reveal many seeds. The valves in drying bend back, bringing pressure to bear on the seeds; and when this is sufficient the seeds split off, being thrown quite violently in all directions.

Unlike the Sweet-scented Violet it cannot be used for flavouring or to decorate expensive chocolates, but its roots have been used in an attempt to cure many things, especially diseases of the skin.

Field Pansy (Heart's-ease) *(plate 49)*
Viola arvensis Murray

Very similar to and hence easily confused with

Wild Pansy (Heart's-ease)
Viola tricolor L.

These are very similar and easily confused. They even share a common name. Whether it was the former, the latter or both, which John Evelyn grew in his garden at Sayes Court near Deptford we do not know. However in 1810 or thereabouts, Lady Mary Bennet made a garden in the shape of a heart on her father's estate at Walton-on-Thames, and planted it with these weeds. William Richardson, her father's gardener, helped her in these endeavours and hybridized the weeds with surprising results. James Lee of Hammersmith seized upon the opportunity and crossed the Walton plants with others from Holland, resulting eventually in our modern garden Pansies. The name is evidently derived from French *pensée* because the face of the flower is bowed as if in thought, and Culpeper says that the wild plant is also 'held to be good for the French Pox', which I suppose might be thought of as a disease from, if not of, the heart.

CLASS 20: ORDER 2
WATER-MILFOIL FAMILY (see also page 183) HALORAGACEAE
Prickly Rhubarb
Gunnera sp.

An introduction from the wet upland forests of South America, its broad rhubarb-like leaves, which may grow to more than 2m (7ft) in diameter, are well-armed with sharp prickles. These gigantic leaves and equally robust spikes of flowers are produced from a massive, part underground, rhizome system; and once it has a hold in your water garden it can take over in a very impressive way.

CLASS 20: ORDER 9
ARUM FAMILY   ARACEAE
Sweet Flag
Acorus calamus L.

Thanks to its sweet smell, it was the favourite strewing rush. Since carpets arrived it has been recorded as growing about the rivers in London, but much reduced in quantity from over-zealous collecting.

Lords and Ladies
Arum maculatum L.

Of all our native plants this probably has more common names than any other, some of which—owing to the phallic nature of the spathe enshrouded in a hood—are commoner than others. So every time it blooms, we have an object lesson concerning the efficacy of the Linnaean two-named system. *Arum maculatum* is much easier than Kitty-run-down-the-lane-jump-up-and-kiss-me.

The root of the plant was used to make a singular pudding called Portland sago, prepared on the Isle of Portland and said to have been more like arrowroot than real sago, which is obtained from the trunk of a tropical palm. The method of preparation must have got rid of the poisonous element, for eaten raw it has a burning taste.

Arum starch was used to give strength to the enormous ruffs worn by Queen Elizabeth I and her subjects. The washerwomen of the time must have rejoiced when the fashion languished in the reign of James I, for the starch blistered their hands without mercy.

CLASS 21: ORDER 1
SPURGE FAMILY   EUPHORBIACEAE

A family whose members usually bleed white sticky latex when they are damaged. Manioc which forms a staple food in the tropics, the Hevea tree which produces much commercial rubber, and the African plant which produces the castor oil which was both a source of and a cure for much discomfort when I was young, are all of this family. Please note, your pet Rubber Plant is not, for it is a kind of Fig.

Petty Spurge
Euphorbia peplus L.

This little spurge bleeds copious white latex which is uncomfortably peppery to the tongue. But though a common weed, it should not be over-maligned for it provides a good organic mulch for the soil. In America, a close relative known as the Gopher Weed because it grows around gopher villages, is actively being re-searched as a future provider of petrochemicals. Trials to date are very promising so here is a plant both to go for and, in the future, go with.

CLASS 21: ORDERS 2 AND 3
SEDGE FAMILY (see also page 135) CYPERACEAE

The members of this family all have their leaves and their flower parts arranged in threes. Their stems too—or at least part of them—are usually triangular in cross-section, and are neither hollow nor swollen at the nodes, thereby distinguishing them from both the rushes and the grasses.

Although very grasslike in appearance and performing

much the same function in providing herbage and cover and binding the soil, this family has not provided us with nearly as many useful plant products as the grasses. Their non-usefulness, for as we have seen no plant can be called useless, accounts for the fact that the taxonomists have left them lumped into few genera, unlike the grasses which appear to have been a splitter's paradise. The importance of the sedges, especially in our upland and our poorer pastures, should not be underestimated. They bind and hold the soil and provide welcome food.

Cyperus and its related genera occupy the warmer parts of the world, while the members of the genus *Carex* take over in the colder regions and indeed become dominant and hence of extreme importance in some of the poorest habitats.

Pond Sedge
Carex acutiformis Ehrh.

Hairy Sedge
Carex hirta L.

Graceful Sedge
Carex acuta L.

Oval Sedge
Carex ovalis Good.

PLANE FAMILY   PLATANACEAE
London Plane *(plate 50)*
Platanus × *hybrida* Brot.

This is a hybrid between at least two other planes, one from America, the other from Europe, and it was first recorded about 1700. If there is any truth in the story that the hybridization took place thanks to John Tradescant's introduction of the American tree into London, then this plant should have the distinction of being called London's only endemic plant. With its hollow leaf stalks on leaves which fall while still green, and a flaking bark which shrugs off the effects of pollution and always looks dappled even on the greyest day, this tree is so suited to city life that it is now planted in towns across the world.

REEDMACE FAMILY   TYPHACEAE
Reedmace *(plate 51)*
Typha latifolia L.

Bulrush to many, which it is not, and Cat-tail to the Americans, this very common plant grows in slow-moving water.

The yellow pollen with fine hairs produced on the upper male part of the spike is so inflammable that it will flash if lit. The down produced lower on the female part of the spike has been used to stuff pillows and bolsters, and Gerard recommends it for blistered heels. A modern use relates to the fact that it can grow in water grossly polluted by chemicals, where it acts as a sort of living sieve.

NETTLE FAMILY   URTICACEAE

Not all members of this family have stinging hairs, but those that do, irritate by producing a complex carbohydrate in the base of a glandular hair tipped by a spine of silica.

Mind-your-own-business
Helxine soleirolii Req.

Another name for this plant which was introduced into our greenhouses and has escaped to the warmer parts of our gardens is Mother-of-thousands, and *Helxine* means 'baby's tears'. It came from Corsica and lives up to its name spreading rapidly given the chance. Once it has got a foothold, selective herbicide may have to be used.

Small Nettle *(plate 52)*
Urtica urens L.

The smaller of our two native nettles, and what it lacks in stature it also lacks in stinging power.

Stinging Nettle *(plate 52)*
Urtica dioica L.

This has been recorded from the Palace Gardens but is no longer on record as still growing there.

Despite their bad reputation, nettles have had many uses including food; and both nettle soup and sandwiches are very good to eat, especially if prepared from young shoots. Pepys in his diary of 1661 states 'we ate nettle porridge which we made on purpose today'. The Romans it is said, planted a third species in Britain for use as a rubefacient. To warm their limbs against the British weather they beat themselves with nettles—a process called urtication, for obvious reasons. The fibres found within the plant have been used in making paper, twine and even cloth. Nettle juice was also used for stopping up fine cracks in wooden tableware and the leaves for packing around fruit en route to market. Rennet can be made from a strong infusion of nettle leaves, and may be kept stabilized with salt until required.

The Common Nettle is an important food plant for many of our insects, including the Tortoiseshell, Red Admiral and Peacock butterflies.

BIRCH FAMILY   BETULACEAE
Silver Birch *(plate 53)*
Betula pendula Roth

Birch
Betula pubescens Ehrh.

Both of our common birches are present, planted in the Palace Gardens. The former has silver-white bark which becomes gnarled and fissured into diamond-like bosses at the base; the latter is usually grey brown and smooth all the way down to ground level. Birches are the hardiest of trees. They grow in the most northerly latitudes, were the first real trees to return after the Ice Age, and thrive on poor acid ground.

In Scotland it is true to say that in the old pre-plantation days everything was made from birch, the commonest wood. Loudon gives the following list: houses, tables, bed, chairs, carts, ploughs, fences, barrows, ropes, dishes, spoons. Its branches were used in the distillation of whisky, and its twigs for smoking herring and ham. The northern variety *Betula pubescens* ssp. *odorata* is especially good, for its twigs are covered with resinous warts which add a very special aroma and hence flavour to both products and to highland woods.

Great untidy masses of twigs called 'witches' brooms' are often seen on the trees. They are caused by an infection which makes the tree believe that it has lost its leader shoot and so it branches again and again. The bosses or burrs from which these branchlets arise are turned into the thinnest of wooden bowls. The bark which comes off with ease in sheets was rolled to become the traditional Yule log of the poor, and the sap can be used to make a refreshing wine.

Most important of all, these trees form an important food source for insects and birds. They are also of great beauty hence in the language of the flowers they signify gracefulness (see page 129).

Alder *(plate 53)*
Alnus glutinosa (L.) Gaertner

A tree which likes to grow part in, part out, of water; its roots which may float free being the home of symbiotic bacteria which form tightly packed woody nodules looking not unlike a clove pomander. These fix nitrogen from the atmosphere and supply both the tree and the soil with nitrate fertilizer.

Charcoal made from the wood was used in the manufacture of black gunpowder, and the inner bark is an explosive vomative when fresh and a purgative when dried. The bark was also an important source of tannin and the wood yielded an orange-yellow dye.

Never try to warm yourself by an alder log; but burning slowly it will keep the fire in all day.

Famous ones at Loughton in Epping Forest still bear the mark of the common rights of pollarding, that is lopping, firewood from living trees. Each year the commoners had to reclaim their right by starting work promptly on 11 November. Each year by guile of too much free drink the night before and eventually the banning of the loppers from the area, the landlords did their best to stop this right of common. In 1866 the commoners of Loughton took on seventeen landlords in the law courts, and they won the day. Of their £7,000 compensation, much went to build Lopper's Hall, a local meeting place.

CLASS 21: ORDER 8
HAZEL FAMILY   CORYLACEAE
Hornbeam *(plate 54)*
Carpinus betulus L.

Possibly the last native tree to reach England after the Ice Age, it still has its only stronghold in the South East. Famous ones at Loughton in Epping Forest like the Alders above still bear the mark of the common right of pollarding.

A hard wood, excellent for burning and also used for the manufacture of mill cogs. It burns like a well-lit candle and was used as such, hence perhaps the name Lant Horn.

Hazel *(plate 54)*
Corylus avellana L.

Our only native tree to produce edible nuts which must have been carried north by birds, animals, water (they float) and by the tree's own invasive growth at the end of the Ice Age. Hazel has been managed as coppice for many centuries and provides poles and small wood for a whole variety of purposes including wattle for daubing and for fencing. Some of the old coppice stools may well be the oldest living trees in Britain for their management can be traced back more than a thousand years.

At last many landowners are beginning to realise the importance of broadleaf woodland in relation to the sporting activities and long-term investment potential of their estates. Coppice with standards may well be the answer in many cases.

BEECH FAMILY   FAGACEAE
Beech *(plate 55)*
Fagus sylvatica L.

Native or not? The argument still goes on for the first real records, both pollen and macroscopic, date to the Bronze and Iron Ages when people were already well-established in all regions of the country. However it got here, it has certainly made itself at home; and producing as it does the densest of canopies and the deepest of shade, it would always win through and become the dominant species if our woods were left to their own devices. It is said that Queen Victoria liked to have logs from Burnham Beeches in the fires at the Palace. It is well known that the wood is highly valued by furniture craftsmen; and until recently chair bodgers used to turn the wood on treadle lathes worked in the open air, their livelihood never outstripping the rate of production of wood by the trees which formed their living workshop.

Sweet Chestnut
Castanea sativa Miller

Introduced by the Romans from southern Europe, its roasting nuts are now as much a part of Christmas on London streets as its handsome presence is part of the countryside.

John Evelyn in the seventeenth century wrote that 'the nut is a tasty and masculine food for rusticks at all times, and of better nourishment than cole or rusty bacon, yae, or beans to boot'. He was so taken with them that he planted them in many places and there is a legend, perhaps based on some truth, that there was a great forest of Oak

Royal Fern
Osmunda regalis L.

(plate 62)

and Chestnut, two miles south of Westminster at a place called Borough Moor.

Pedunculate (Common) Oak *(plate 56)*
Quercus robur L.

What more need be said? It provides a home for more than 340 species of insect, it did the same for the people of medieval Britain and long after, and gave them a Navy to defend their shores.

Turkey Oak
Quercus cerris L.

Introduced from Europe, it has bristly acorn cups and provides a home for some of the Common Oak's many tenants.

The Lucombe Oak is a natural hybrid between the Turkey and the Cork Oaks, and was first raised in Exeter in 1765 by a nurseryman of that name. It is a delightful tree whose old leaves fall just as the new ones open, it is thus sub-evergreen and is planted widely as an ornamental.

WATER-MILFOIL FAMILY (see also page 179) HALORAGACEAE
Spiked Water-milfoil
Myriophyllum spicatum L.

A native plant with finely divided leaves which are each made up of 13–35 segments. It grows submerged in, and floating on, water, and lives up to its name, blocking millponds and streams and getting wound up on water-wheels, causing many a wheelwright to say wrong things. This otherwise delightful little plant provides food for waterfowl and cover for a great variety of aquatic organisms upon which our game fish depend.

CLASS 21: ORDER 9
Scots Pine *(plate 57)*
Pinus sylvestris L.

A magnificent tree, but in its true Scots form of *Pinus sylvestris* ssp. *scotica* (Schott) E. F. Warb. it is the essence of that northern principality. The uses of the tree are manifold. Its straight-grained wood itself is used for many purposes, and the resin which flows so freely from any wounds was the basis of a whole chemical industry.

Pine torches, made by drying the trunks of saplings which have been split into many strips, in an oven, give off a good white light and continue to burn even in a strong highland wind. 'Wool' fibre made from the leaves was used to stuff cushions, pillows, mattresses, and even seats in railway carriages. The pleasant smell given off by this pine-cotton proved beneficial in many ways. The bark of the tree was not only used for tanning but the inner bark as flour for bread and the wood chips instead of hops in beer.

All in all, a capital tree: tall in growth when young, and in age its spreading crown and pink-washed branches give open shelter to a great variety of plants which includes the shy Twin-flower, *Linnaea borealis* itself.

CLASS 21: ORDER 10
CUCUMBER FAMILY CUCURBITACEAE
Melon
Cucumis sp.

A family which is represented in our native flora by White Bryony and in the menu of Royal Garden Parties by the form *Cucumis sativa* L. In America a whole variety were grown by the Red Indians and later by the settlers. In 1961 a seedling of a melon turned up in the Palace Gardens but it was never identified—a fact which is enough to give anyone the pip. In a good summer melons may be grown outdoors and can run riot over your garden and those of your neighbours, so take care.

CLASS 22: ORDERS 2, 3, 4 and 5
White Willow *(plate 58)*
Salix alba L.

Goat Willow *(plate 58)*
Salix caprea L.

Common Sallow
Salix cinerea L.

There are more than 300 species of willow in the world and no less than 22 in Britain, many of which hybridize with each other, making identification a job of devotion.

Many of them are also very variable and are split into a number of sub-species. For example, the Common Willow growing in the Palace Gardens has red-brown hairs on the underside of the leaf mainly on the veins and is the sub-species *atrocinerea* (Brot.) Silva & Sobr. If it took three expert botanists to name it, you can see the problem and may even perhaps ask, 'why bother?' Well, the best cricket bats are made from *Salix alba* var *coerulea* (Sm.) Sm.—that is why we bother, for each type of willow has its own very special properties. All of them, however, produce nectar and feed insects.

Willow bark and wood have long been used to cure headaches, and they work. The active principle is indeed the chemical blueprint from which aspirin (acetyl-salicylic acid) was developed.

CLASS 22: ORDER 7
White Poplar
Populus alba L.

Lombardy Poplar
Populus nigra L. var *italica* Muench

The former, with leaves which are white beneath, is an introduction. The latter, the famous Lombardy poplar, though closely related to our native Black Poplar, was introduced—being a native perhaps of Central Asia.

These rapidly growing trees produce an abundance of wood which does not splinter and hence is ideal for matchsticks, punnets and Camembert cheese boxes. They also produce the best white wood for firing bakers' ovens, and are said to help make the best French bread.

CLASS 22: ORDER 8
Dog's Mercury
Mercurialis perennis L.

Although members of the Spurge Family neither of the mercurys bleed white latex, but both turn blue black very quickly when they are damaged. One might think this species is called Mercury because it comes up so quickly in spring, or because it is poisonous to man and beast alike; but no, it is named after the God who, it is said, discovered its virtues.

Theophrastus relates of *Mercurialis perennis* 'that if a woman use to eat either the male or female mercury, two or three days after conception, she shall bring forth a Child either male or female, according to the sex of the Herb she eats'. That is, if she lives, for the plant is very poisonous. Another name for it is Town Weed, for it is common in urban places.

Annual Mercury
Mercurialis annua L.

The hairless branched stems and annual habit, serve to distinguish this species from the last. It is not poisonous, the leaves being eaten in Germany as spinach. The leaves are rich in mucilage but again are not a source of latex.

CLASS 22: ORDER 10
SIMAROUBA FAMILY SIMAROUBACEAE

An exotic tropical-sounding family, the members of which are found mainly in the tropics.

Tree of Heaven
Ailanthus altissima (Miller) Swingle

Introduced to Britain from China to give autumn colour in the garden. Growing very quickly, it reaches up to the sky, hence its common name.

Its flowers are either unisexual or bisexual and it is strange that only those plants with male flowers smell, and in this case that is the correct word for it is an offensive odour. Rub the leaves and find out which sex is present. Today it is well established; its fruits have wings which, being slightly twisted, corkscrew down.

Older readers may have come across another member of this family in the form of quassia chips which when rubbed upon the thumbs of adolescents prevented them from being sucked—a very bitter remedy and memory.

CLASS 22: ORDER 13
CONE-BEARING PLANTS GYMNOSPERMAE
Yew *(plate 59)*
Taxus baccata L.

No true Englishman could ever call this noble tree a weed, for it gave strength to the long bows which both defended our shores and fostered our ambitions abroad. It is poignant to note that no less than three of our kings—Harold, William Rufus and Richard Coeur de Lion—

bowed to the supple strength of yew and that yew of foreign origin was held in greater esteem by our bowmen.

To offset this inestimable virtue, the leaves of the plant are poisonous to cattle and extracts of the same have, it is said, been used like those of the Foxglove to affect the heart.

A common tree within the walls of churchyards where it was safe from stock, and the stock were safe from it, it has always been held in some awe. The supposed great age, circa 2,000 years, of some churchyard specimens, indicates that the superstitions attached to it may predate the Christian era.

The bright coral fruits are certainly not poisonous to birds, nor to country children who eat their slimy substance under the name of Snottygoggles. The bitter hard seed contained within however is, and its being spat out by the children and passed through the birds may account for the appearance of yew seedlings as weeds in untended corners of churchyards and palaces alike.

CLASS 23: ORDER 1
GOOSEFOOT FAMILY CHENOPODIACEAE
Common Orache
Atriplex patula L.

Halberd-leaved Orache
Atriplex hastata L.

Members of the Goosefoot Family (see also page 144), these two plants, though commonest near the sea, are equally at home in the open spaces of farms and gardens especially where there is manuring or other fertilization. This tolerance or almost dependence on high levels of nutrients probably stems from the natural habitat of these plants just above the strandline where high tides and waves wash in and deposit all manner of organic matter. Likewise these plants are tolerant of high concentrations of salts within their chosen habitat. A plant with a leaf shaped like a beefeater's halberd is a fitting plant for any palace garden.

The family is of great importance having provided us with beetroot, spinach and the much maligned mangold-wurzel whose name means 'root of scarcity'.

As you will see from their placing by Linnaeus, goosefoots have perfect flowers while those of the oraches are unisexual.

CLASS 23: ORDER 2
HOLLY FAMILY   AQUIFOLIACEAE
Holly *(plate 60)*
Ilex aquifolium L.

Holly is holy, the red fruits at Christmastide signifying Christ's blood and the flower buds which become cruciform as they open in May echo the Crucifixion. It is also, along with Ivy, our commonest evergreen native plant. A latecomer into Britain after the Ice Age for it cannot tolerate prolonged frost, it existed as an undercover plant until polished stone axes began to change the sylvan scene. Then it came out of hiding to flower and fruit along the edges of the expanding fields, protected from the stock by its prickly inedible leaves.

CLASS 24
CRYPTOGAMIA

It must be remembered that at the time when Linnaeus was developing his sexual classification, little or nothing was known concerning the microscopic details of the process of reproduction in plants. Hence he grouped all those plants which did not make public their reproductive strategy by means of a flower and fruit, into this final Class of 'hidden marriage'.

CLASS 24: ORDER 1
FERNS AND THEIR KIN
PTERIDOPHYTA

The very word 'fern' conjures up a picture of a delicate, flimsy, filmy thing, one of a shy and retiring group of plants existing in shady retreats—not at all the sort of plant to be found in the list of Royal Weeds. Perhaps the stark facts that fern-like plants were abundant upon the earth for almost 150 million years before the first true flower bloomed, and that they are all perennials with efficient underground rhizome systems, is enough to explain the presence of at least some in these lists.

Common Horsetail *(plate 61)*
Equisetum arvense L.

Of all the Royal Weeds, this plant boasts the longest lineage of all, for its ancestry goes back almost 300 million years, when the remains of giant horsetails fell into tropical swamps and so laid down the vast deposits of coal upon which the Industrial Revolution and the wealth of the British Empire was in part founded.

During the reign of Queen Elizabeth I it was sold by the herbe-women of Chepeside under the names of Shave Grass, Pewter-wort or Vitraria, for the purpose of cleaning the wooden ware or treen of the common table and burnishing the pewter which garnished the tables of the upper classes. Common horsetail owes its name to its appearance—like the tail of a horse; and its mild abrasiveness to quantities of purest silica—vegetable glass—laid down within its tissues.

One of the most difficult weeds to eradicate; but disliking shade cast by other plants it may be eliminated by overplanting with Nasturtiums.

Royal Fern *(plate 62)*
Osmunda regalis L.

From the common to the Royal, this fern was once widespread across the British Isles, revelling in damp acid places especially in the west. It was indeed so common in places that the farmers of the Lake District were in the habit of covering their potatoes with its fronds to guard them from frost en route to market. Habitat destruction and perhaps over-collecting have now made it a rare plant in its natural setting, but if given a chance it can act like a weed.

Osmunda was one of the titles of the god Thor; *Os* = house; and *mund*, as in Sigismund and Edmund, signifies strength and power. The herbalists used the word Mundyfye and said that it gave strength to the human system. Certainly the rhizome and roots when boiled become very slimy and were used in northern Europe to give strength (stiffening) to linen.

Bracken, Common Brake or Eagle Fern *(plate 63)*
Pteridium aquilinum (L.) Kuhn

Undoubtedly the commonest of all our British Ferns and a real weed, for though it shuns lime-rich soils it enjoys the company of man, and thrives from many of his practices. Perhaps because it is so common, it is linked with many stories and superstitions, not the least being how it got its name *aquilinum* from Latin *aquila* (eagle).

Cut the leaf stalk aslant just at the point where it enters the ground and you will see the Prussian Double Eagle picked out dark against a lighter background. However, if you cut the stalk straight across—a perfect section—you will see an Oak tree. It is said that the latter Royal Oak only appeared in the stalk after such a tree gave shelter to King Charles.

As to its supposed virtues, the strangest alludes to its spores, although in early accounts they were of course called seeds. If the seed is gathered as it becomes visible on St John's Eve, it can be used to make you become invisible. Whether this strange virtue stems from the fact that of all the ferns this has the most inconspicuous sporangia, or that the Duke of Monmouth hid himself after the Battle of Sedgemoor in a clump of bracken, I do not know.

Hart's-tongue Fern *(plate 64)*
Phyllitis scolopendrium L. Newman

With its strap-shaped undivided leaf, this is the easiest of our ferns to recognise. Though a native of woodland ravines especially in the west, it is often found growing in drains, even in the middle of cities. It was indeed in such a gutteral habitat that it was found on the West Terrace of the Palace Gardens in 1961.

Its preference for such dank habitats and the way it gets to the more obscure localities depends on its spores. Unlike the Bracken, the spore-bearing structures of this fern are very conspicuous. Each leaf has some 80 in number, each containing 4,500 sporangia, each of which produce 50 spores. A prodigious 18 million spores per leaf and there are 10 leaves to an average plant each of which can, if it reaches the right habitat, grow into a scale-like prothallus upon which are borne the organs of sexual reproduction. These facts were not known to Linnaeus, hence he used the term Cryptogamia, hidden marriage, the product of this hidden union being a new hart's-tongue. Its uses and virtues were much acclaimed by the ancients, but little has been upheld.

Bracken
Pteridium aquilinum (L.) Kuhn

(plate 63)

Wall Rue
Asplenium ruta-muraria L.

Again its presence on a shady Palace Wall is due to the abundance of light spores which characterises this group. The leaves of the plant resemble in outline those of the Garden Rue, but may immediately be told apart simply by looking at the veins. The veins in this, like in all lowly plants, are dichotomously branched, that is at each branch point they divide into two equal halves—a characteristic not found in the higher plants.

Old walls have long provided a habitat for many plants, especially where the mortar provides lime in what is otherwise an acidic environment. If the Queen has never seen it in her London home, it is abundant on the crags of Arthur's Seat behind her Palace in Edinburgh. Though usually a small plant it was held in high renown as a cure for coughs, probably because it yields a little mucilage when boiled.

Male Fern
Dryopteris filix-mas (L.) Schott

Ranking alongside the Bracken as our second most common fern, it is again found at the centre of many tales and legends. The term male, which is found in many countries—Feli Maschia in Italy, Polypodio Helecho Masculino in Spain, Fougère in France—appears to have its origins in the robust nature of this fern when compared to the more delicate Lady Fern with which it often grows.

The esteem in which its supposed virtues were held is perhaps best summarised by the names it was given—Lucky Hands, St John's Hands, Johannis Wurtzel; for it was thought that anyone in possession of such a treasure was safe from all ills, including witchcraft.

Common Buckler Fern
Dryopteris dilatata (Hoffm.) A. H. Gray

Even the young coiled form—which in strict botanical terms is referred to as circinately vernated—is characteristic of the group and resembles the top of a bishop's crozier. Whether the resemblance relates to the mystic powers of each or either, I do not know. However, while still in this young state the leaves of this species may be distinguished from those of the Male Fern by the numerous broad scales which help to protect them. In this species the scales have a dark, almost jet-black centre.

Another feature of distinction is that the underground rhizome of the buckler fern does not creep and spread, but grows upwards; producing in old age almost a trunk which may be 50cm (20in) tall.

Wall Rue (*Asplenium ruta-muraria*)

A weed is but a plant whose virtues have not yet been discovered. All I can now suggest is that you judge each plant for itself and, as you do, think upon the following frightful facts.

The plant kingdom comprises some 380,000 plus species. The plus must be a large number, for to date the vast majority of the earth—and especially the tropics which are the seat of much of that diversity—has not yet been adequately explored by botanists. We live in a world society which reaches out to other planets but has not as yet bothered to catalogue all the plants which form the green core of the support systems of space-ship Earth.

At the present rate of destruction of natural and semi-natural vegetation, reliably estimated at 150 acres per *minute*, the world is in danger of losing its plants, many of which we cannot even put a name to, let alone list their virtues. One in ten of the known species are at this moment in danger of extinction. Even those which are not in immediate danger are having their genetic diversity, and hence their chance of long-term survival, cut back

(left) Toad Rush (*Juncus bufonius*), (centre) Slender Rush (*Juncus tenuis*), (right) Field Wood-rush (*Luzula campestris*)

day by day—and that includes our crop and garden plants. Genetic uniformity has never been a virtue in any plant, in fact it is exactly the opposite; for the more uniform a plant population, the less chance it has of survival.

To date but a tiny percentage of even the plants we can put names to have been studied in relation to their virtues, their potential use for food, pharmaceuticals—some 40 per cent of the world's drugs come from natural sources—timber, pulp, oils, resins, waxes, insecticides and other chemical raw materials, let alone their role in soil stabilization, formation and regeneration. If any are ever proved to be without virtues, then and only then, should they be looked upon as weeds.

So, I believe the definition holds and we must all wait with hope for their virtues to be discovered. It is indeed the only hope we have, for once all the coal, oil and natural gas—the fossil fuels which were laid down by 'weeds' of the past are gone—we will have to turn more and more to the diversity of the plant kingdom to provide the organic raw materials of our chemico-plastic way of life. We cannot, however, wait that long, for many of our country's weeds are no longer as common as they were.

In 1945 there was as much broad-leaved and coppice woodland in Britain as there had been in Middle Ages—a diversity of ancient unimproved pasture fed a sufficiency of at least dairy cattle. Village ponds, fens and wetlands were a source of local wonder and a home for a multitude of fish, waterfowl and plant and insect life. Our hedgerows, some dating back beyond Saxon times, bound the living landscape in good heart and provided real farmers with an abundance of game birds, and their wives and children and workers with beautiful surroundings in which to live out their hard-working lives. Despite a world war, our world-renowned heritage landscapes were then still intact, and once more began to bring in multi-millions of tourist revenue to enjoy this green and pleasant land.

This is no longer true; forty years has seen much of this swept away. The uniformity of cereal and conifers has taken the place of the diverse productive beauty of coppice. Butter and barley mountains and milk lakes have replaced more than 90 per cent of our 'unimproved' grasslands, which used to be self-sustaining, but now require the constant attention of machinery, and the whole incredibly expensive chemical armoury of agriculture. Our wetlands and our hedgerows are in tatters, our common heritage almost gone.

Natives and visitors alike have in fact been more and more alienated from the free enjoyment of that heritage. These once common plants are no longer common in our countryside, except where they are held in trust by caring people. God save the Queen and all her once Common Plants.

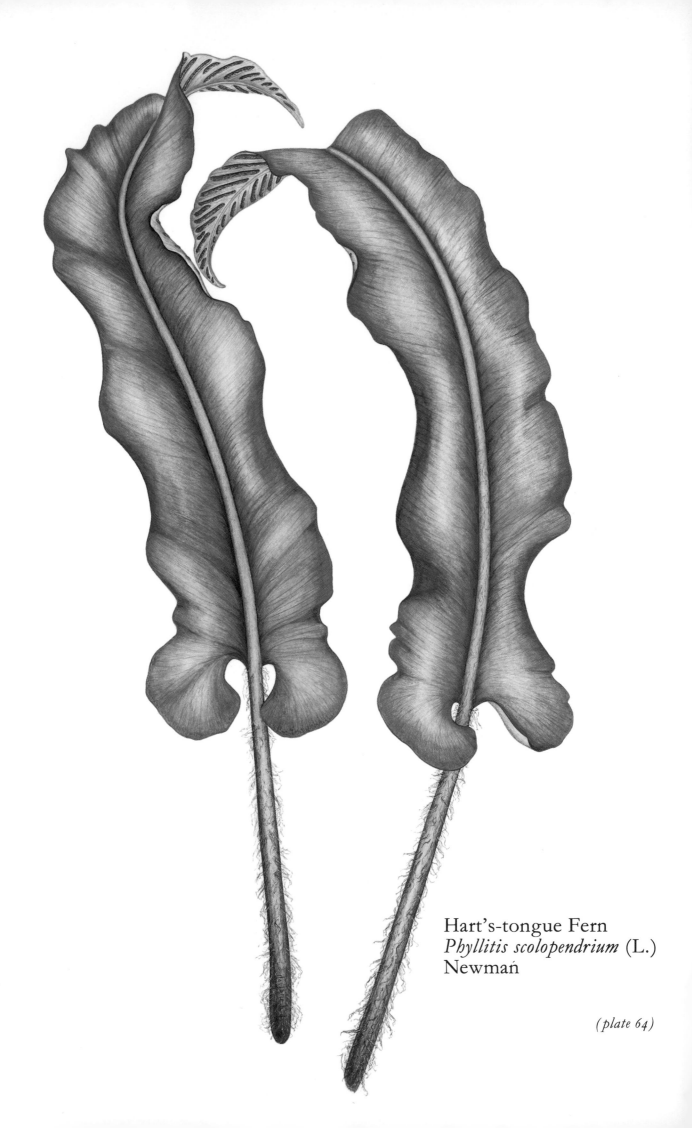

Hart's-tongue Fern
Phyllitis scolopendrium (L.)
Newman

(plate 64)

THE FAIRY STORY WHICH
CAME TRUE

Very early on, when writing this book, I came across a problem. The records set down in the *Proceedings and Transactions* of The South London Entomological and Natural History Society were very detailed. As you may guess from the society's title, the section concerning insects was simply crawling (and creeping) with information. No less than 343 species of Lepidoptera (butterflies and moths) had been put on record up to 15 September 1964, a staggering 14 per cent of the then known *British List*, which had itself been upped by two—for two insects new to the British Isles had been found in the survey, *Monochroa hornigi*, a moth which until then had been known to be living around and within Knotweeds in Austria and Germany, and *Earias biplaga*, known as a pest of the cotton crop across Africa.

Fascinating information for dedicated entomologists, but not in itself enough reason to designate the area of the Palace Gardens a National Nature Reserve, nor even a Site of Special Scientific Interest. However, add to these and all the other records (which again to most of us are just long Latin names) fascinating facts like the following concerning *Tinea columbariella*, and all of a sudden these become more than just records of mere insects:

> ... the larva of which inhibits the nests of pigeons, feeding on their debris. This species was until a decade ago confused with *T. pellionella* (the Case-bearing Clothes Moth) but unlike its congener is seldom encountered. Two of its strongholds in England are the towers of the British Museum (Natural History, of course) and Canterbury Cathedral, where it thrives as a symbiont of the pigeon population of these buildings.

Fascinating and very important, for like so many of the insects and other smaller animals in the record books many have an important function to perform as natural refuse-disposal operatives. Take them or the plants on which they depend away from the garden, and these natural processes which cleanse the soil and help re-cycle the mineral nutrients grind to a halt. Their place can only be taken, and not very satisfactorily, by expenditure of more human time, perspiration and money on mechanical devices and fertilizers.

In the bad old days, London smog—the only marginal advantage of which was that it gave London some really spectacular sunrises and sunsets—all but banished evergreen conifers from our streets. It also perhaps kept back tar spot from our Roses and the Dutch disease from our Elms, but these marginal pluses were by far outweighed by the minuses which concerned the health of both the public and the plants. The soot coated our lives and our lungs as it coated any leaf which dared to last from one year until the next through the smog-bound days of autumn. A covering of soot meant less light could reach the all-important chlorophyll, and so even the most magnificent trees just ticked over or died where they stood.

Partly in consequence of this there were no conifers of any size in the Palace Gardens at the time of the survey, yet the following species of moth were recorded; Larch Pug, Grey Pine Carpet and five which have no English names. But the larvae of all of these feed on conifers, and we can count ourselves lucky because they may well be an indication that the Clean Air Act has really made a difference to all our lives. The Queen's gardeners can count themselves lucky that after the final count it seems safe to say that 10 per cent of all the country's species of insects are probably there, most of them helping with the garden chores. Not all, I must agree, for some are real nuisances.

The vast majority of insects are of great beauty and/or interest if only we bother to take a look and find out, and are completely dependant upon the treasury of common plants either as a food source or in some other way to help them complete their complex life cycles. In the same way they, as we have seen, help the plants to complete their life cycles, so providing us with fruits and seeds and the potential of a crop next year.

This then was my problem—how could I do full justice to the labours of those entomologists and to the subjects of their interest, without which the whole living world would begin to grind to a halt?

On Friday 8 June 1983 I travelled to London by train, filling much scrap paper en route with my deliberations concerning this point. I was on my way to the Isle of Dogs to attend the annual prize giving at St Luke's Church of England Primary School. It was a fabulously festive occasion, complete with a wealth of entertainment by the children and with mums, dads, grandmums, grand-dads, aunts, uncles and teachers crowding the very hot school hall to applaud both the prize winners and the show.

The highlight of the entire programme, at least for me, was a tableau complete with little people dressed as flowers, creepy-crawlies, butterflies and moths, depicting what has become a very traditional tale about a garden. It was also very appropriate, for the climax of the proceedings was to be the opening of the school's own Wildlife Garden, complete with pond, beds of aromatic flowers and banks of weeds—the whole due to the vision and sheer hardwork of the pupils, staff, parents and the local authority. We all poured, glowing, perspiring and sweating, out from the heat of the hall across even hotter tarmacadam to cut the ribbon (to one side of course so that it could be used again), and entered their own garden of delights. It was a super warm day and the garden was alive with insects, not gi-normous, gaudy ones, but small ones each doing its own thing. Where were most of them? You've guessed—around the weeds.

My problem was solved. Their tableau and their garden had said it all. So at this point I decided to do no more than to indulge myself in my life's third great interest. The first two are botany and children, the order depending on what has happened in the garden or at the teatable on any particular day—the third is ballet. I am very proud of my FLS (Fellowship of the Linnaean Society), and equally proud that I am one of the patrons of the West Midlands Youth Ballet, who work out of Solihull and perform in the fantastic auditorium at the University of Warwick. The end of 1984 being the deadline for publication of this book, and 1985 being the tenth anniversary of the ballet company, everything came together and a true-life fairy story began.

To me, ballet is the perfect amalgam of the sciences and the arts, coupled through a hundred per cent fit human body. Music, movement, musculature can all be regarded, can all be analysed, through science; and are in the dance synthesised into the perfection of the arts. I can think of no better way of doing justice to this subject than through a ballet, inspired by children, to be performed by young people of great dedication, whose aim in life is the seeking of perfection.

Lords and Ladies (*Arum maculatum*)

COMMON HERITAGE

(A Ballet in Three Acts)

Set along the Great Border of the Gardens of Buckingham Palace

ANIMALIA **THE CREEPY CRAWLIES—BUTTERFLIES AND MOTHS**
Holly Blue (*Celastrina argiolus*)
Large White (*Pieris brassicae*)
Swallow-tailed Moth (*Ourapteryx sambucaria*)
White Ermine (*Spilosoma lubricipeda*)
The Herald (*Scoliopteryx libatrix*)
Garden Tiger (*Arctia caja*)
Red Admiral (*Vanessa atalanta*)

PLANTAE **THE QUEEN'S FLOWERS**   **THE QUEEN'S WEEDS**
Camellia 'Elizabeth'   Holly
Sweet-scented Stock   Shepherd's Purse
Lilac   Privet
Delphinium 'Royalist'   Dandelion
Rose 'Queen Elizabeth'   Willow
Lilium regale   Dock

PERSONAE Her Majesty Queen Elizabeth II
Garden-party Goers

Swallow-tailed Moth (*Ourapteryx sambucaria*) and caterpillar on Privet (*Ligustrum vulgare*)

ACT I

It is a warm, dry, wilting day, and the caterpillars are hungry—oh, so hungry. They approach the flowers growing in the Great Border and enquire if they may feed from their leaves. The flowers rebuke them, saying 'No! We are the Queen's, and we are ready for the Royal Garden Party. We are far too important to be eaten by mere insects, nasty creepy crawlies.'

ACT II

It is evening on the same day in a less well-kept corner of the flower bed. The caterpillars are now desperate, for without food they will die. Each in turn asks one of the weeds for sustenance, which is willingly given.

A Red Admiral butterfly joins the party, asking no more than a place in which to lay her single green egg. She tries each weed in turn, but none will suffice for her life cycle can only be completed on a Stinging Nettle, all of which have been banished from the garden. Heartbroken, she dies.

ACT III

It is the day of the Garden Party. The Queen and her guests arrive to be enthralled by the beauty of Holly Blue, Large White, Swallow-tailed Moth, White Ermine, The Herald and Garden Tiger—two butterflies and four moths appearing in all their summer glory as they cast away their caterpillar coats while dancing in the breeze.

A fairy story, yes, but it is all based on fact. What is more it is a fairy story which came true. Malcolm Williamson CBE, Master of the Queen's Music—an Australian who is not only one of the world's foremost musicians, but a great conservationist who did much to support the Tasmanian Wilderness Society in their fight to stop the despoliation of a World Heritage Site—agreed to write the music.

It is also a fairy story which can come true right in your own back yard.

BELLAMY'S BOOK OF WILD GARDEN ETIQUETTE

THE GOLDEN RULES

1 Never remove any plant from the countryside to plant in your garden. For a start in many cases there are laws to help prevent you, and for a finish why should you steal them? Leave them there for others to enjoy and as a food source for insects, birds and animals.

The same is of course true for botanic gardens, stately homes, municipal parks, next-door neighbours and the like. It amazes me that people who care enough to want something special to grow in their garden do not seem to spare a thought for all the effort that the owner gardener has expended to grow it in the first place. It's stealing, and often so stupid, because 'them as asks very often gets', and with it gets good advice on how to grow it. The end result of most hit and run jobs is a dead specimen in the furtive depths of your handbag or coat pocket.

This is even more true when you are on holiday abroad. There you are a visitor and, coming from the country which engendered not only the birth of democracy but also of conservation, you should always set a good example, however tempted. Also remember that it is unlawful to bring propagatable material of any sort into the country from abroad, even from the EEC, without an import licence. These laws are not there to give HM Customs more to do, but to protect our gardens, crops and forests from disease. Please, if you are on holiday with a group which includes sponge-bagging plant collectors, ridicule them for what they are—vandals and idiots.

Sleeping Beauty (*Oxalis corniculata*)

2   This rule may seem very harsh, but unfortunately it is true at this juncture in our landscape's history: do not pick wild flowers. Yes, it is harsh! And I must confess that when I was young I liked to pick a bunch of wild flowers for my Mum. That was almost fifty years ago, and today although mums are just as deserving, our wild countryside has shrunk almost out of all recognition. There are just not enough flowers left to go round—at the moment. However, we can change all that with a little bit of garden etiquette.

HOW TO PRODUCE
AN ELIZABETHAN GARDEN

First you must decide exactly what you want from your garden, and exactly how much diversity you can pack in, although this will in the main depend on size. To overcome any argument we will take half an ordinary sized back-garden plot—a definition which is usually a lot longer than it is wide, however you will get the meaning.

The suggestion is that you do all the formal bit—flowers, fruit and veg, greenhouse, patio and potting shed—in the part closest to the house. Of course, if you are a fanatic then you may want to reverse the whole thing or even let the whole garden go wild. But that is the wrong term, for that is exactly what is not going to happen. Remember the joke about the proud gardener and the vicar. 'Well, my son, what wonderful things you and God have done with the garden.' 'You should have seen the mess he made of it on his own.' The natural processes of regeneration will take over your garden, if you let them; it will however take an awfully long time (in gardening terms that is) for it to produce much of great beauty or utility either to yourself or the local wildlife. What we are about to do is help nature to help both herself and you in the garden.

While on the subject of local wildlife, what about the neighbours? If he or she has got fence to fence fitted prize plants of whatever sort, he or she isn't going to take too kindly to whatever you do. If you grow bigger and better prize plants, you will be in one sort of wrong. If you suddenly let your wild plants and animals affect their bit of the local environment, all hell may well be let loose, and it should.

Beating the Bounds

Fences cost a lot to build and maintain, they also produce dense shade for you, and them next door. They however do provide a screen for both the wind, and for weed seeds and, if you aren't too liberal with the creosote, a good habitat for wood-rotting fungi.

High walls have many of the same pluses and minuses, though without the problems or pleasures of fungi. Their pointing, if done in not too much of the gin and tonny style, provides a welcome lime-rich habitat for intra mural lime-loving plants called calcicoles.

Hedges need clipping, and although I like Privet for it feeds nice sorts of caterpillars and I grew up surrounded by it, a bit of variety is a great thing both for you and the animals. However have care, for though hips and haws are fine for the birds, prickles and thorns are appalling, even though the thickest Marigold gloves.

If you can choose, the following plants could go into your well-laid hedge—Hawthorn, Alder Buckthorn, Guelder Rose, Spindle, Wayfaring Tree.

Upper Crust and Social Structure

Now to the main structure of the plot. If you want birds—what a silly thing to say, of course you want birds!—you must build in three dimension and provide somewhere for them to hide. You may well like gawping at them through binoculars, but they do, especially at certain times of the year, like a little bit of privacy. Remember too, that all birds don't fly off to Torremolinos and beyond for the winter. The ones which stay or arrive at that time need leafy shelter, so include some evergreens and conifers amongst your high-rise shrubs and trees. That is exactly what each tree is, a high-rise home for a multitude of life, providing both an upper crust and a pecking order.

When it comes to trees, please be careful. An English Oak may be the grandest and the best in that it harbours both your patriotic sentiments and more types of creepy crawlies than any other type of tree. However, an Oak at the back of a two up two down anywhere could become much more than cause for local aggro. If you do manage to persuade yourself and your neighbours that that is what you all want (and please, I am in no way knocking the idea) I suggest you plant it a long way from your foundations and invest in some of those little notices stating 'Ancient Lights' while they are still going cheap.

Whatever tree or trees you do decide to plant, remember that the sun, when it shines, does so from the south and that your neighbour's plants have just as much right to photosynthesise as yours.

To the trees may be added all sorts of nest-boxes, but these will be dealt with later.

Compost Corner

Some gardens inherit compost heaps, some gardeners learn that compost heaps are (apart from themselves) the most important thing in the garden, while others have compost heaps thrust upon them by the need to get rid of garden waste, by the laws on bonfires, the Clean Air Act and the rising cost of fertilizer.

Please don't get the idea that I don't believe in fertilizers and in garden and farm chemicals. I can and do say with my hand on my heart that without the correct use of such chemicals, world agriculture would be in an even worse mess than it is, and mass starvation would be on a much greater scale. Also, thanks to the pressures put on the market by the conservation bodies, the producers have responded (and spent enormous amounts of money doing it) to produce slow-release fertilizers and safer chemicals of all sorts. Did you know that there is one which only kills Couch Grass and nothing else—absolutely fantastic. All you have to do is read the labels, and in these days of garden centres with built-in fully upholstered patio chairs and tea or coffee (free in some places), you can sit in glorious garden surroundings and do just that. So there is no excuse, bio-degradable, safe, systemic, natural, they are the 'in words', and if you do your bit and only purchase the safe ones, by your natural selection (for that is still the only market force that anyone takes any notice of), the other sorts will go to the wall of extinction even faster.

Garden Tiger (*Arctia caja*) caterpillar and moth
on Clustered Dock (*Rumex conglomeratus*)

All that said, and I hope done, your compost heap should be given pride of place, not tucked away in the darkest coldest corner. It only works, recycling your own and your garden's waste—and don't stint the tea leaves or the burnt chips—thanks to multimillions of bacteria and fungi. Bless their little cells, they don't have warm blood to keep them up to reaction temperature and, working flat out for your benefit, they need and deserve every bit of help they can get. So be proud of your compost corner, be it of the traditional heap variety or a highly sophisticated plastic rotary-cylinder job. Join the club, give it a chance, and it will repay you ten mulching-fold and keep up with everything you sling its way. You can easily check up which is the warmest spot by doing a frost watch in the winter. Check where the frost lies longest, that's the coldest spot; where it disappears first is the warmest. Please note, we are talking about frost not snow; the longevity of the latter will provide slightly different information, for instance where there is a barrier to the wind or where the children built their snowperson. I trust you get the drift of the argument.

Such microclimatic information can also help you choose where to put your other garden plants, and here a little bit of knowledge about where they originated will help; remember King James's Mulberries.

A word of warning and a word of comfort. If the warmest spot does turn out to be close to the house, I suggest you don't put the compost heap there because even the best maintained may hum a little both with aromatics and with flies in the summer. Also put it on the up-wind (prevailing, that is, and checkable from weather socks hung up to dry) side of the garden from your stroppiest neighbour.

Though your compost corner must centre on the heap itself, the environs are of great importance. Back or flank it with Stinging Nettles. This will stop people fuelling it from the wrong side and will provide a food plant for Peacock, Small Tortoiseshell, Red Admiral butterflies and the weird Snout Moth, to name but a few. Plant a Buddleja nearby and make it really easy for the adults to have their tea, as well as having not too far to go to lay their eggs. Remember the poor old Red Admiral has flown in all the way from Europe just to make your garden more beautiful. Give 'im a real welcome and a helping hand.

The Nettles will thrive on the effluvia from your compost, for they are phosphophiles, that is they like lots of phosphate around their roots. So if you are caught short down at the bottom of the garden this is the place to go, but please not in front of the butterflies.

In case your neighbours or any of your family are sundowner-drinkies-in-the-garden addicts and compostophobes, which are usually two of a kind, a good idea is to plant Night-scented Stocks flanking the Nettles. These will not only provide overpowering aromatics at the right time, but a habitat for lots more beautiful insects; but more of that below.

A word of comfort relates to the product from this unique symbiosis between you and all those millions of bacteria and fungi. If you have an active vegetable garden growing on the side, you will probably be able to use up all the good black humus-rich compost that you can ever produce. The same is true if you are into flowers and do a lot of potting, whether it is out or in. However, if you do produce a surfeit of this valuable soil conditioner, offer it to your neighbours or friends, or flog bags of it at the church fair; I have seen more bizarre things on sale. If you do in the end have to put some in the dustbin, remember it will have been much reduced in volume, probably by as much as 50 to 1 or more, and it may well give your local refuse disposal operative management a little twinge of conscience! You see, when they have filled up all your local holes, old quarries, wet fields, little valleys (you know, the places where you used to take your kids and the dogs for a walk), that is exactly what they are going to do with the rubbish—turn it into compost, recycle it, and so reduce one problem of our effluent society down to more manageable proportions. Meanwhile you can start by doing your bit and improve your garden, its soil, and your balance of payments into the bargain.

The Serpentine Lake

No wildlife garden is complete without a stretch of water. It may be a lake or just a permanent puddle, but whatever the size it will have a life all its own. I can guarantee that wherever you live in Britain you can enjoy the shimmer of Damsel-and even Dragonflies' wings at the most a couple of years after letting the water in.

You will also have your own whole *Life on Earth* cross-section of evolution, minus David Attenborough (although you might persuade him to pop round) living in or visiting it. To see many of these water creatures—the plankton, necton, seston and so on—you will need a microscope or at least a good magnifying glass. However, some of them are of larger proportions, and frogs, toads and newts all need a place to mate and lay their spawn, and it needn't be very big. Create a pond and some may well get there under their own steam. However, you can help the process by giving a home to those frog-, toad- or newt-poles (the latter will be good company for any friend who falls in after cocktails) which need a home at the end of the Easter term. Please, please, please don't collect amphibia from the wild; all are getting rarer and the newts are now on the endangered and protected list. If you want spawn, poles or even adults, there are suppliers; or you could ask your local members of the Royal Society for Nature Conservation (silly me, you are a member) for help from one of the many schemes which are now spawning all over the country. If you want a permanent colony of amphibians, you will need a deep patch of water with some stones on the bottom for winter

Spiked Water-milfoil (*Myriophyllum spicatum*)

201

hibernation, also some stones sticking out at the edge or a launching ramp, for metamorphosis. The frogs will then be able both to drop in and hop out. Talking of dropping in, there are a lot of animals that will do just that. Hedgehogs for a drink after their bread and milk, martins, swallows and bats to help keep down the insect life, herons after the goldfish (a good tip here is to warm the water and keep piranhas, but if you do, don't tell the RSPB). You can even adopt a duck, through the Wildfowl Trust's scheme, and hope that one day as it migrates it will call in for a visit. Even if yours does not come to pay its respects, others may well find your pond is just what they need for a stopover. Then if you are really keen, you can let the local primary school or Girl Guide troop in for a dip. No, it hasn't got to be that deep, I mean a pond-dip with nets. Not only they but you will be surprised at what they find, but please, if it is alive, make sure they put it back. With diligence your wet corner could become the focal centre not only for your local natural, but also for your local human, community.

How to build the thing? Well, this is not the place to go into all the pros and ponds of construction, there are plenty of books on the subject. But don't make it too formal, and the best way I know of starting an informal pond is a pond-puddling party. All you need is a nice warm day, a hole with sloping sides already dug, a hosepipe and a load of clay. There are suppliers of such super gooey stuff all over the country.

The best attire is an old bathing costume, shirt and bare feet—in the latter case, the younger the better. The order of the party is to soak the clay and all jump in, and puddle. This is a verb which has its very wet roots in the Old English to 'clart' and 'splodge' and describes a process akin to treading grapes. It feels much nicer too, no pips or skins to get stuck between your toes—just the sheer heaven of puddled clay oozing amongst the piggies. The fascinating thing is that as the afternoon wears on, the task gets easier and easier as the clay gets squidgier and squidgier. The end result is that the whole party gets puddled and you have a hole complete with a watertight liner. If I may be so bold as to offer advice on the catering side, I would suggest strawberries, cream and a Sauterne 1961, for though the job is good fun like the wine it is tiring. Black polythene is much cheaper than clay, but nowhere as much fun, even if you are into that sort of thing.

Finally, please remember the edges, for at least one margin of the pond should be puddled back to provide a marsh or swamp. There you may develop your own hydrosere complete with King-cups, Gipsywort, Water Mints, Greater and Lesser Spearworts, Yellow and Purple Loosestrife, Skullcaps and many more waterside plants, all of which will bring in the insects. This too is the place for the gnome with the fishing rod, each one guaranteed to send you postcards when he has been stolen away on holiday.

Never, if you can help it, put your pond under trees. The plants which will grow in it are already shaded by the water and they therefore need all the light they can get. The plant plankton also need light and warmth, and the earlier they can get cracking in the year the more food there will be for the other inhabitants. Also too much of even a good thing can be a killer, and autumn leaves falling into the pond will clog up the whole works, and will make the pond anaerobic. This may preserve the pollen so that future generations of palynologists can pry into your garden record, but it won't do the majority of the pond-life any good. So site your pond away and up wind from the trees, and if any leaves do fall in rake them out, putting them on the compost heap. If you are bashful, do it on a moonlit night, especially if you live in Wiltshire.

The source of water is also another important consideration. In summer the level will go down due to evaporation, and the shallower the pond the worse will be the consequences. You must do your best to keep it topped up, but the use of tap-water can cause problems of its own. If there is a drought, use of the garden hose may well be restricted, a rule to which you should adhere. Even if you can use it, tap-water may be rich in nutrients with other additives like chlorine and fluoride. Think what too much of the latter might do to your piranhas' ring of confidence.

Joking aside, such additives and especially too much in the way of nutrients will upset the balance of life in the pond, which will start to grow green slime and go very smelly. So have a care and, if you can, use a rainwater butt as a source of supply. Such a reservoir fed from the roof of the house, garage or even garden shed will itself make a delightful oasis for microscopic water life and a boon for watering the garden or even washing your hair; for it is soft, mineral-poor, almost-distilled water.

One interesting feature of the fauna of waterbutts is the microscopic Water Bears or Tardigrades. Under the microscope they look not unlike a little 'Michelin man' with eight stubby legs each ending in a claw. They are neither insects nor spiders, but belong to a half-way sort of group all their own. A missing link which can't be missing because it is living in your gutters and, if you give it a chance, in your water butt. Better use tap-water for washing your hair after all.

The Greensward

'It's your turn to mow the lawn'—or at least what is left of it amongst the moss and the Daisies—is a weekly cry in our house throughout the growing season. And usually someone (for extra pocket money) grumblingly goes and gets out the motor mower which drinks up even more pocket money and shatters the peace of the garden. I hasten to add that our garden had more than an acre of the stuff, all well-clipped until we learned the secret of that man and his dog who went to mow a meadow, but only once or twice a year.

Again it is decision time. You must decide the overall use and hence regimen for your lawn space. Part or parts of it, either in the middle of the main expanse or more likely a section that abuts onto fence, hedge, shrub or tree which gives dense shade, can be developed as a meadow. Not just full of green grass and Daisies but with a blue, pink, red, white or even bright-golden haze of flowers, fruits and seeds just before mowing time. All hay fever sufferers should note that such a decision isn't all bad, for although the grasses will produce pollen, the more showy flowers there are, all of which are insect pollinated and so chuck less of the offending stuff about, the less grass there will be.

The real beauty of it is in the mixture of plants. Sorrels, Knapweeds, Scabious, Dog Daisy, Corn-flowers, all growing nice and tall at the back. Then Cat's Ears, Cowslips, Hawkweeds, Harebells, Lady's Bedstraw, Lesser Stitchwort, Sheep's-bit, Yarrow, Yellow Rattle, growing less tall in the mid-ground; with Daisy, Eyebright, Lamb's Lettuce, Self-heal, Speedwell, even Moonwort, growing as little 'uns at the front.

Better still, have three different patches—they needn't be very big, a few square feet will do—each one of which is allowed to grow to a different height, being mown or cut in different ways and at different times of the year. Real estate-management stuff. Your eighth of an acre suddenly takes on new dimensions, it has Windsor Great Park capabilities. The short meadow should be mown regularly, but with the blades never set below $1\frac{1}{2}$in. The medium-height area should be left until early July each year and then cut regularly. The tall should be left until September (lucky you) and then cut when all the late summer flowers have set their seed. This is very important because what you are aiming at is a self-propagating area, or areas, of lasting beauty.

The rules of the meadow game are:

1 Keep the rest of your lawn well mown or you will groan as the meadow takes over the lot.
2 Always remove the clippings (the seeds will have fallen off) to the compost heap, for within reason the poorer the soil is allowed to become, the better will be the crop of flowers. Use a scythe or sickle and, when you do, practise first a long way away from the children. The money you save in petrol in the machine, you can give to the World Wildlife Fund.
3 Never use fertilizer or lawn chemicals on the meadow parts of your estate, although regular soil ventilation using a fork will help all parts of the lawn.
4 Perennial Rye Grass is too aggressive for most of the meadow flowers. If you are starting a new lawn from scratch, never plant this variety. If it is there already, the low-nutrient regime will help to eradicate it as will the addition of seeds of nice grasses like Common Bent, Doddering Dillies, Sweet Vernal, Crested Dog's Tail and Yorkshire Fog.

You may well decide to do away with all forms of mechanical help and use rotational grazing by animals such as tortoises, rabbits or guinea pigs. All can be moved in at the right time under open-bottomed runs. Never tether tortoises by the shell, it is ultra-cruel. Any such animal husbandry is not as good as the sickle or scythe, because the animals don't remove their clippings—sorry, droppings. Perhaps it is best to make hay while the sun shines and feed it to the animals in another part of the garden at another season of the year.

ADVICE TO YOUNG MOTHERS AND
OTHER LEPIDOPTERISTS

The Queen has 343 + species of butterflies and moths which live in and visit her garden. The vast majority, all except five, are in fact moths and most of those are small. However each one has its own detailed beauty and intricate lifestyle, the study of which can be of absorbing interest. The vast majority also have an important part to play in the processes which are vital to the running of your garden, from pollination to the breakdown and decay of litter.

What is more, they have in the main evolved to feed on or live in our native wild plants, and many do little if any harm to the cultivated ones. I am not saying that there are no insect pests, unfortunately there are, too many of them for anyone's liking. And though a Leaf-cutter Bee may be of great interest to watch, if it nibbles your prize rose leaves just before show time, well that's enough to send anyone rushing for the spray-can. However most bad attacks are from something specific and should be treated in a specific way (ask at the garden centre). Don't waste money and gall by using wrong and blanket treatment; to kill the lot is stupidity in the extreme.

You may ask why bother to have all the weeds and their eaters there anyway? The answer is that they form the nucleus of a whole natural management system made up of plant eaters, meat eaters, predators, dung eaters, refuse-waste and litter eaters. If you kill all of those off not only do you kill the baddies but with 'em go the goodies as well. Goodies, like the Ladybeetles—Ladybirds to most. How many spots have yours got—two or seven? You haven't counted! Well they are easy enough to see for these, our most brightly coloured common insects, are marked in a startling way to warn the birds that they taste real bad. Their bright

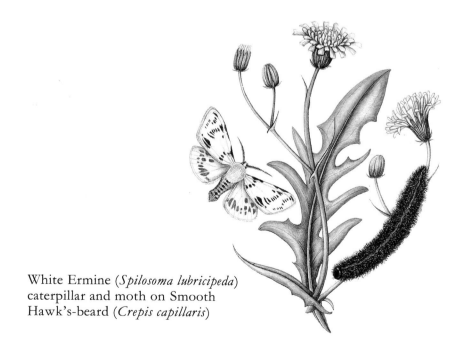

White Ermine (*Spilosoma lubricipeda*)
caterpillar and moth on Smooth
Hawk's-beard (*Crepis capillaris*)

coloration doesn't however warn off the Blackfly and Greenfly upon which they feed voraciously—kill off the Ladybeetles at your roses' peril. I've got a 22-spot in my garden: thank goodness at last manufacturers are putting such bright markings on their garden products to warn gardeners of their good and bad properties. All the time we are learning from nature.

If I now say that butterflies can't do any harm, every vegetable gardener will run screaming until he turns Cabbage White. What is more there are three of them to choose from—Large, Small and Green-veined White. Which do you hate most?

The female, and let us stick with the Large White which has two black spots and a black stripe on the top side of each of her fore-wings, lays conical yellow eggs in batches on the leaves of cabbages and their kin in late spring. They hatch, and a few caterpillars start nibbling at the leaves. Not much to worry about but, if you want to, this is the time to attack. When they have had sufficient, they crawl away and pupate on fences and the like, some being cheeky enough to come into your garden shed where they stick themselves on the wall with a silk safety-belt—'clunk click', they are up to every trick. Out come the adults to feast on nectar pollinating your Buddlejas and Petunias, along with thousands of others who have migrated in for their gastronomic holidays to the Continent. They all then lay their eggs, and later the cabbages really start to suffer; for unlike most human children, the caterpillars eat up all their greens. Beautiful the butterflies may be, but they are real baddies in the vegetable garden and they must be dealt with.

So what about the goodies. It is a fact that many of this second brood of caterpillars will never complete their life cycle for they have been parasitized by another sort of insect, an Ichneumon Fly. She lays her eggs in the caterpillars, perhaps a hundred or even more in each. They hatch and the grubs give the baddy real cause for concern, eating it from the inside out, muscles first and vital organs last, so that it lingers on as a walking snack bar. Yes, it's all happening in your garden, and though you may say 'Yuk, horrible', it is probably no worse for the caterpillar than a dose of the wrong insecticide. It's just one of the very many interactions which help keep the natural processes in control if you only give them a chance. Destroy that little Ichneumon Fly and Britain's cabbage patches could be overrun in a season, Dolls and all.

So all in all I reckon it is worth making a place for the creepy crawlies and enjoying those we can, like the larger butterflies and moths. Here follows a list of plants, both weeds and cultivated, that may help to give you a first in the Butterfly-gardeners' Lists. C and A stand for something very famous in the world of adornment. Here C stands for Caterpillar for caterpillars use the plants as a source of food and shelter. The adults, at least the females, have to visit the plants in order to lay their eggs which themselves come in a whole variety of beautiful shapes and colours, single or massed. A therefore stands for Adult, for the flowers of these plants attract them in order to sup nectar. So make sure your garden has its fair share of C and A plants and you can rest assured that when the time comes you will have your fair share of adornment from both visiting and resident butterflies and moths. My advice to lepidopterists, young and old, is to plant a Buddleja for a start and buy a book and keep your own unique garden record.

Spring

Bird's-foot Trefoil	C	Silver-spotted Skipper
Honesty	C	Orange Tip
Milkmaid	C	Orange Tip
Wild Strawberry	C	Grizzled Skipper
Anemone blanda		
Aubretia		
Heathers and Heaths	A	Important to many which have just come out of hibernation
Yellow Alyssum		

Summer

African Marigolds (large)	A	Tortoiseshell and Golden Moth
Bird's-foot Trefoil	C	Silver-spotted Skipper
Broom	C	Silver-studded Blue
Buckthorn	C	Brimstone
Cowslip	C	Duke of Burgundy's Fritillary
Cow Wheat	C	Heath Fritillary
Devil's-bit Scabious	C	Marsh Fritillary
Dock	C	Small Copper
Gorse	A	Silver-studded Blue
Ground Ivy	A	Checkered Skipper
Honeysuckle (wild)	A	White Admiral, Elephant Hawk-moth
Hop	A	Comma
Horseshoe Vetch	C	Adonis Blue
Lavender	A	The dreaded Whites
Oxford Ragwort	C	Cinnebar Moth
Parsley Family	C	Swallowtail
Pea Flower Family	C	Clouded Yellow
Privet	A	Privet Hawk-moth and many others
Restharrow	C	Common Blue
Sorrel	C	Small Copper
Thistles	C	Painted Lady
Vetch	C	Small Blue
Violets	C	Fritillaries

Autumn

Dahlias	A	Brimstone, Peacock, Painted Lady, Red Admiral, Tortoiseshell
Fallen Fruit	A	Red Admiral
Hebe	A	Comma, Red Admiral, Tortoiseshell
Horseshoe Vetch	C	Chalk-hill Blue
Ice Plant	A	Tortoiseshell
Michaelmas Daisy	A	As for Dahlias
Pea Flower Family	C	Clouded Yellow
Yellow Knapweed	A	Brimstone

Two Bursts of Activity—May and Late Summer

Cabbage Flower Family C	Whites
Holly C	Holly Blue
Ivy C	Holly Blue
Nettles C	Small Tortoiseshell, Peacock

Winter

Ivy Flowers A	Comma, Red Admiral, Tortoiseshell

As regards obtaining the plants, they can be obtained from any good nurseryman; ask advice and he should be able to help. The golden rule is to usually go for the oldest varieties wherever possible, for they very often appear to be the most attractive to insects, being full of old-time aromatic magic.

Wild plants can be bought from good seedsmen and there are more and more coming onto the market all the time. However, there is one great drawback with these, for if you use them you will be planting non-local stock in your locale. That cannot be looked upon as the best thing to happen for it will dilute the local special genes. If there is no other way, go ahead; and if you can, choose a mixture of packets from a number of seedsmen. That doesn't guarantee genetic diversity, but it is a step in the right direction.

There is another way which is much more fun, but please have great care. Remembering the golden rules at the start of this chapter, and that wherever you go in the countryside you have big feet, plan to set up your own local gene bank. Patience and green fingers will help.

Locate your nearest source of the common wild flowers you want—note the word common; and never use a local nature reserve, country park etc. Visit these flowers during the fruiting season and remove no more than one or a few seeds from a number of the plants, and the more you move about (with great care) the better will be your chance of sampling genetic diversity.

Plant the seeds out, just as you would those you buy at the garden centre; a bit of trial and error will be required for they don't come complete with instructions. Treat them with care, prick them out, harden them off and plant them out in your butterfly borders, moth margins, mini-meadows, semi-detached hedgerows, copses and compost corners. Then sit back and enjoy the spectacle year after year after year. Never, but never, be tempted to take them out of your garden and plant them back into the natural and semi-natural vegetation of the real countryside. That is best left to get on by itself although it can often do with a little bit of protection, so please make sure you are a member of your County Conservation Trust, and the Friends of the Earth.

You can, of course, spread the good word of natural gardening to your neighbours and swap seeds and even caterpillars. What is more you can have butterfly-watching parties, which are just as exciting as listening to the dawn chorus and better in some ways because you don't have to get up so early. Did you know that many butterflies are territorial and will guard their territory, even against you. Watch them drink, unrolling those fabulous long 'tongues'; watch them mate and lay their eggs, see how careful each Mum is. Then there are the antics of the caterpillars—gnawers, gnashers, loopers, mimics, danglers, leaf rollers. They will go on enchanting you with their own special ways of life long after the adults have gone and of course before they come back again. How do they manage to squeeze all that beauty into the dull ugly pupal case and then unfurl and pump up their gorgeous scaly wings, and where do they do it? The answer is— right in your own garden.

Water Plantain (*Alisma plantago-aquatica*)

If you have a stroppy next-door neighbour who doesn't like the birds nicking his buds, point out that other birds help keep down the largest insects and suggest that he or she keeps a cat or puts nets over the prize crop. Real gardeners with their eyes on a good crop will do that anyway. If the neighbours start to complain that your weed seeds are about to blow their way, gently turn a hose on them—the seeds that is, not your neighbours—and wash them off. Best of all, when there is something fantastic like a nest full of baby birds, a Damsel-fly emerging, or two Slugs mating suspended up in their own world of ardour, invite the neighbours round to see it and share in the knowledge and so the excitement. You could even lend them a few Ladybeetles when the Aphids are winning.

Hairy Sedge (*Carex hirta*) and
Bulrush (*Schoenoplectus lacustris*)

CREEPY CRAWLIES
AND THE GARDEN SHED

The garden shed may be the nerve centre of all your hard work where tools are put away spotless each night and where your chemical warfare arsenal is stored; for at some time even the most efficient natural gardener may have to resort to force and push the button. Of course it is stored safely away from the gaze of little eyes and hands in the proper containers properly labelled. Better still if it is always kept in one corner and in a locked cupboard, but even the best gardeners are not all angels.

Although operational decisions may be made over a cup of tea in the kitchen, the garden shed is mission control, where the action starts. It is however also a haven of rest, not just for Dahlia tubers and other non-hardy propagules hanging it up through the winter, but also for a cross-section of our garden wildlife. There is great excitement, even a visit from the local newspaper (and so there should be) when a robin nests in the old kettle in the corner, but there is so much more which goes unrecognised, unenjoyed and hence unplanned for.

First and foremost there are the spiders, for it is here the really big ones may be found. Do note that they have eight legs and a two-bit body and are therefore not insects. Note also that none of the British species can bite or poison even the most delicate of us human beings. If you are a spiderphobe you won't believe a word of it, but it is nevertheless true. So if you can stand them, enjoy them, and the garden shed is a good place to start.

Check the webs, particularly the orb webs around the edge of the window, shimmering in the sun and catching all the tiny insects as they try to get in. They probably belong to *Zygiella* who, though she may not be obvious on the web, is there in a silken tubular retreat waiting for something to drop in. Then between the boards or planks or, if you have an upmarket shed, between the bricks—wherever there is a convenient chink—you may well find little almost cottony webs with a hole in the middle looking just as if they have been stuck on the wall. These are made by one of the wall spiders, *Amaurobius*, which is quite big and sits there picking off wandering creatures, some of which can do your shed and your gardening year a lot of harm. The real cobwebs in the corners produce those large disturbing spiders which dart out to catch everything unawares—the true House Spiders, with bodies up to $\frac{3}{4}$in long, called *Tegenaria*.

But there are very many other spiders, mainly out in the garden doing a Trojan job of keeping the other creeping crawling life down to proper proportions. It has been estimated by the great arachnophile W. S. Bristowe, that all the spiders in the British Isles together eat the weight of its human population in insects each year—the only reason I can think of for not slimming. There are about six hundred different sorts of spiders feeding in the British Isles; the list for Buckingham Palace tops sixty. It was first studied in May 1929 when a morning's search yielded twenty-nine species, and Bristowe records a letter from Clive Wingham which speaks of King George V's delight on such a good catch. One interesting spider that has emerged in the records of the palace cellars is a smaller relation of the

Daddy-long-legs spider, appropriately this one has a blue body. I have searched my garden shed for it in vain, for alas it is an import in the straw used to pack cases of good French wine. I wonder if it lives inside those less-exalted wine containers between the cardboard and the silver-foil bladder? I wonder too whether James I would not have had more success if he had tried to produce spiders' silk.

(left) Trailing St John's Wort (*Hypericum humifusum*), (centre) Common St John's Wort (*Hypericum perforatum*), (right) Square-stalked St John's Wort (*Hypericum tetrapterum*)

Other fascinating denizens of the shed are our overwintering butterflies. Please leave the top window open in the autumn to let them in, and again when the spring warms up to let them out, and keep the environs of the top window-frame clear of spiders' webs. A bit of creepy crawly ecological planning can help make the shed a more perfect place for everything. Let the web builders have free range in the height of summer, that is after all when the little insects are at their most annoying. At the end of summer clear the webs by brush, if not by hand, from all top corners, the roof and the top half of the windows. Leave them everywhere else. Keep the top window open and in the butterflies will come, Small Tortoiseshells and Red Admirals, to hibernate for the winter, perched, if that is the right word, on the ceiling in the corner; that's why it is best to remove the cobwebs in that vicinity. If one of them does perch too close to a web, you can very gently pluck it off, holding its wings together between finger and thumb and pop it in a safer place, having patience while it latches on again. The best thing, however, is to leave well alone, just remembering to let the butterflies out in the spring. The Small Tortoiseshell will want to get up and go on a warm day in January, so be sure there is a nice crop of Ivy flowers waiting for it.

It would be fascinating to know what sorts of insects flew in the Palace Gardens back in the days when Dryden wrote his long poem of adulation. Were some of our butterflies and moths which have now become rare due to habitat destruction and the wanton use of insecticide there in vast numbers? Unfortunately we have no way of knowing.

We do however know that in the Duke of Buckingham's time there was, at the end of the greenhouse, a little 'closet of books' in which his favourite volumes were arranged for ease of reference, perhaps including Robert Sharrock's *History of the Propagation and Improvement of Vegetables by the concurrence of Art and Nature* and described as:

> Shewing the several ways for the Propagation of Plants usually cultivated in England as they are increased by Seed, Offsets, Suckers, Truncheons, Cuttings, Slips and Circumposition, the several ways of Graftings and Inoculations, as likewise the methods for Improvement and best Culture of Field, Orchard and Garden Plants, the means used for remedy of Annoyances incident to them; with the effect of Nature, and her manner of working upon the several Endeavours and Operations of the Artist.

This was published in 1660 and was for long the handbook of gardening. A copy is one of the great treasures of my own library and on page 136 it states:

> Boggy Plants require, even when they be planted into Garden, either a natural or an artificial Bog, or to be placed near some water, by which there is great improvement to all sorts of Flags, and particularly I have observed to Calamus Aromaticus.
>
> The artificial Bog is made by digging a hole in any stiffe clay, and filling it with earth taken from any Bog, or in want of such clay ground, there may be stiffe Clay likewise brought in, and laid to line the hole or put in the bottom or floor and the sides, likewise.

(above) Dog's Mercury (*Mercurialis perennis*) and (below) Petty Spurge (*Euphorbia peplus*)

You see, they were at it back in those days—both gardening with the help of nature, and puddling.

The book dwells much upon the insects, mostly bad. Dealing with earwigs he writes:

> Setting Beasts Hoofs among the Flowers, upon sticks, is used of every Body here to take them, and generally lik'd. Some that set their Flowers in Pots, set the Pots in Earthen Plates, with double verges, containing water or water mingled with soot in the outer verge, to drown the Vermine that shall attempt the pots, and rainwater in the second, which may pass through the holes in the pots to water the earth therein contained.

This is chemical warfare and hydroponics all in one.

If Buckingham did have this book in his ordered library, I am sure he would have gained much fascination from the creepy crawlies. For under the windows of his ordered closet was a little wilderness full of Blackbirds and Nightingales, and I am sure lots of insects.

May I suggest that you should likewise have a store of books ordered and ready to hand, and I implore that one of these should be Michael Chinery's fabulous *The Natural History of the Garden*, which will provide many hours of happy reading and a whole lifetime of happy hunting around the environs of the garden shed.

UP THE GARDEN WALL

In the absence of a real garden wall do not fret. Everyone has got a house and so a wall at one end or the other. And even if it is made of wood it comes within the category of a wall for the purposes of this our penultimate discourse on garden etiquette. Every such wall has chinks, some are large and put there on purpose so that we may blink through from behind the chintz to view our handiwork or that next door. Others, usually smaller, provide habitats and footholds for all sorts of things; a state of extra-mural affairs which can be improved on, thus bringing the wildlife into close perspective.

The upper windowsills of Sir Alec Douglas-Home's house—The Hirsel, near Coldstream, Berwickshire—are each adorned with a Swift-box. The occupants of his home can therefore have the fantastic pleasure of leaning out and opening the lid to see the proud parent at close quarters. This rare pleasure can only be enjoyed during the breeding season for these most active birds fly the rest of their lives away, feeding on flying insects and sleeping on the wing. They only come to rest when they are nesting. It is not just the Swift-boxes but the whole garden and estate which is a credit to the name of Douglas-Home—a family which has included a number of excellent natural historians among its members.

If you don't want Swift-boxes, a good window-box will bring wildlife close, as will a bird-feeding table or one of those plastic fish-net stockings full of nuts. The pedestal variety of bird table is best, especially if you are a cat lover as well. And remember to provide water for drinking and bathing and a good supply of good food; but no salted nuts or ready salted crisps please, however much you give them to drink. Also stop feeding during the breeding season, so that the mums and dads can wean the chicks on their natural diet. Then when they fly off to another territory where the gardener isn't quite so kind, they will be able to fend for themselves.

If your wall faces south, you might like to add a bat-box. These delightful creatures, a number of which are endangered species, and all of which are protected by law, may come in and roost in the summer. Remember they roost upside down, so provide a 'perch' near the top of the box. If you move the box to another wall so that it does not get too hot on warm winter days, they may even like you enough to join you for hibernation.

There are the other sorts of nest-boxes which come in all sizes and monstrous shapes at the local garden centre. Before making up your mind the best thing to do is join, or drop a note to, the Royal Society for the Protection of Birds—they are the experts. There are plenty to choose from: Tit-boxes, Woodpecker-boxes, Robin- and Flycatcher-boxes and even boxes for House Martins. Whichever you choose, and especially if you choose the latter, site it where the nest rubbish will not be a nuisance. Birds are very house proud—of their house not yours—and will empty the contents over the side where it can annoy humans or damage prize plants.

Nest-boxes should always be put in the shade, in a position where they cannot be reached by cats and can be watched without disturbance. If you do have success with your Swift-box, don't lift the lid too many times or too rapidly. Finally you should clean the boxes out and disinfect them at the end of the breeding season, not just as the next one is about to start.

Creeping Cinquefoil (*Potentilla × italica*) on left and Silverweed (*Potentilla anserina*)

Last, but by no means least, don't forget to provide covering for the wall itself. Native plants include Honeysuckle, Old Man's Beard and Ivy. All provide food, and all except the latter give good cover for birds. Horticultural varieties of plants are legion, among them the Pyracanthuses and Cotoneasters provide beautiful colour and welcome food in the autumn. Climbing Roses, ever so English, provide hips and spiny protection from cats. Quinces and Jasmine are marvellous in spring; Clemati and Virginia Creeper do their beautiful best throughout the year.

Mind you, they do need management and, if left, will not only hide the wildlife but will make new chinks in your wall and eventually even cover the big ones. So, it's best of luck.

CREEPING UP WITH THE JONESES

Never set out to beggar your neighbour, all the way along the best policy is to share your expertise, your plants, but above all your experiences. If your garden does have more beautiful and interesting creepy crawlies than the one next door, don't be ashamed of a little feeling of pride; however, don't brag about your good fortune, share it. I can and always will get excited when a new garden plant comes onto the horticultural scene. I will, however, be just as excited when my Gillyflowers, Dame's Violets, Lady's Smocks and Eglantine attract a butterfly to

White Mustard (*Sinapsis alba*) and Wild Radish (*Raphanus raphanistrum*)

lay her eggs, a bee to seek out nectar or a Spotted Flycatcher to catch a visiting fly. Perhaps more excited; for they have been doing it since long before there were gardens and garden weeds.

There are more than a million acres of gardens in Britain, most of them owned by ordinary people like you and me. With the mass destruction of our countryside for whatever reason—and this is not the place to argue the pros and cons of agricultural policies whether common or rare, although I would at the moment like to know who is not being conned—conservation must start at home. And you can see that it does, there in your own back yard. Your natural gardens will, I hope, give our country's wild flowers and wildlife a chance until the powers to be come to their senses and realise that a little bit of natural etiquette in the countryside is the only way for real long-term meaningful investment.

Go on, have a go! You will never better the cucumber sandwiches or the tea, but when it comes to your own back-garden nature reserve, at least it is worth trying to keep up with the Windsors.

BIBLIOGRAPHY

ALLAN, Mea. *The Gardener's Book of Weeds* (Macdonald & Jane's, 1978)

BARKER, C. A., MOXEY, P. A. and OXFORD, Patricia M. *Woodland Continuity and Change in Epping Forest* (Field Studies 1978, 4, pp 645–669)

BURTON, Rodney M. *Flora of the London Area* (London Natural History Society, 1983)

CHINERY, Michael. *The Natural History of the Garden* (Fontana/Collins, 1977)

CLAPHAM, A. R., TUTIN, T. G. and WARBURG, E. F. *Flora of the British Isles* (Cambridge University Press, 1962)

COATS, Peter. *The Gardens of Buckingham Palace* (Michael Joseph, 1978)

CULPEPER, Nicholas. *Pharmacopoeia Londinensis* (George Sawbridge at the Bible on Ludgate Hill, 1675)

DEALLER, Stephen. *Wild Flowers for the Garden* (Batsford, 1977)

DEAS, Lizzie. *Flower Favourites* (George Allen, London, 1898)

DE CRESPIGNY. *A New London Flora* (Hardwicke & Bogue, London, 1877)

GIRLING, Maureen and GRIEG, James. *Palaecological Investigations of a Site at Hampstead Heath, London* (Nature Vol 268, 7 July 1977, pp 45–47)

HARPER, J. L. *Population Biology of Plants* (Academic Press, 1977)

HUTCHINSON, J. *Evolution and Phylogeny of Flowering Plants* (Academic Press, 1969)

KNIGHT, David. *Ordering the World: A History of Classifying Man* (Andre Deutsch, 1981)

LEE, James. *An Introduction to Botany* (London, 1788)

LEYEL, Mrs Carl. *Green Medicine* (Faber & Faber, 1952)

NORDENSKIÖLD. *The History of Biology* (Tudor Publishing Co, New York, 1928)

PHILLIPS, Henry. *Floral Emblems* (Saunders & Otley, London, 1825)

PRATT, Anne, revised by Edward Step. *Flowering Plants, Grasses, Sedges and Ferns of Great Britain*, 4 vols (Frederick Warne, 1905)

RACKHAM, Oliver. *Trees and Woodlands in the British Landscape* (J. M. Dent, 1976)

RHODE, Eleanour. *The Old English Herbals* (Longman, London, 1922)

SCAIFE, Robert G. Several excellent typewritten reports

SHARROCK, Robert. *The History of the Propagation and Improvement of Vegetables, By the concurrence of Art and Nature* (A. Lichfield for Thomas Robinson, 1660)

Proceedings and Transactions of the South London Entomological and Natural History Society (1963) *Natural History of the Gardens of Buckingham Palace*

STEP, Edward. *Herbs of Healing: A Book of British Simples* (Hutchinson & Co. New York, 1926)

WARD, Cyril. *Royal Gardens* (1912)

WILKINSON Gerald. *Trees in the Wild* (Bartholomew, 1978)

INDEX

Illustrations are indicated by *italic* type